Rock and Mineral Magnetism

W. O'REILLY, BSc, PhD, MInstP

Senior Lecturer in Geophysics and Planetary Physics
University of Newcastle upon Tyne

Blackie

Glasgow and London
Distributed in the USA by
Chapman and Hall
New York

Blackie & Son Limited,
Bishopbriggs, Glasgow G64 2NZ

Furnival House, 14–18 High Holborn, London WC1V 6BX

Distributed in the USA by
Chapman and Hall
in association with Methuen, Inc.
733 Third Avenue, New York, N.Y. 10017

First published 1984

British Library Cataloguing in Publication Data

O'Reilly, W.
Rock and mineral magnetism.
1. Rocks—Magnetic properties
I. Title
538'.7 QE431.6.M3
ISBN 0-216-91460-4

Library of Congress Cataloging in Publication Data

O'Reilly, W.
Rock and mineral magnetism.
Includes bibliographies.
1. Rocks—Magnetic properties. I. Title.
QE431.6.M3073 1984 552'.06 83-20012
ISBN 0-412-00401-1 (Chapman and Hall)

Photosetting by Thomson Press (India) Limited, New Delhi
Printed in Great Britain by Bell & Bain Ltd

Preface

The past two decades have witnessed a revolution in the earth sciences. The quantitative, instrument-based measurements and physical models of geophysics, together with advances in technology, have radically transformed the way in which the Earth, and especially its crust, is described. The study of the magnetism of the rocks of the Earth's crust has played a major part in this transformation. Rocks, or more specifically their constituent magnetic minerals, can be regarded as a measuring instrument provided by nature, which can be employed in the service of the earth sciences. Thus magnetic minerals are a recording magnetometer; a goniometer or protractor, recording the directions of flows, fields and forces; a clock; a recording thermometer; a position recorder; a strain gauge; an instrument for geological surveying; a tracer in climatology and hydrology; a tool in petrology. No instrument is linear, or free from noise and systematic errors, and the performance of nature's instrument must be assessed and certified. This has been the task of the research worker in rock and mineral magnetism. For many workers in the field, and this includes myself, the study of the magnetism of rocks and minerals has been not only a handmaiden in the service of the earth sciences, but has provided the intellectual rewards and scientific challenges of a subject in its own right. The geophysical backcloth against which the problems of technology and the application of the scientific method are met, and sometimes solved, add extra interest, stimulus and satisfaction to that which might be gained in pursuing the equivalent 'pure' or fundamental science.

As in any field of scientific research, so in rock and mineral magnetism there are rival schools of thought in which the merits of one, as yet untested, physical model describing some phenomenon, are promoted against other, untested, models. Even when models are tested, as they must be testable to be 'scientific', the results may not be definitive. The models are invariably simplifications of the real word in which the experiments take place. The results of similar experiments may be in poor agreement, experimental methods and conditions can be criticized, and so on. It is the complications, contradictions and controversies which stimulate and intrigue the research worker. How dull research would be if the advance of knowledge was made in a well-ordained predictable procession!

To write a monograph on a subject might seem to imply that that subject had entered into the canon of scientific knowledge; that the questions posed had been answered to almost universal satisfaction, at least for the time being. Rock and mineral magnetism has yet to reach such old age, and work in the field has, in the phrase beloved of thesis-writing research students, 'raised almost as many questions as it has answered'. It has, however, reached sufficient maturity that it should be possible to write a pleasing account of the basis on which its practitioners operate. Popper has described science as the 'art of systematic oversimplification'. I hope that in preparing this book, in concentrating on what seems to be known or generally accepted, rather than on what is not, I have not given an over-simplified picture of rock magnetism. The scope of the book may with justification be described as narrow in that little or nothing is included on the impact that the magnetism of rocks has had on geomagnetism, tectonics, and planetary physics. This position seems analogous to that of the engineer developing magnetic recording systems who is indifferent to whether his product eventually carries binary codes, Buxtehude or Bessy Smith. Perhaps for the dedicated rock magneticist the medium really *is* the message!

This book carries no formal dedication. Like everyone working in the field I have been the beneficiary of the pioneers of rock magnetism theory and practice—the names of Néel and Nagata come immediately to mind. My insight into the subject also derives from the work of numerous individuals and I hope they, in return, will find something of interest here. The rock magnetism novice was principally in my mind while writing the book.

My thanks go to Susan Banks and Dorothy Cooper, respectively my typist and draughtsperson. I also want to record my thanks to the publishers who provided the stimulus and encouragement without which this book would not have been started, or finished.

W.O'R.

Notation

Where the same symbol is used for more than one quantity, the relevant usage will be clear from the context. The individual identities of members of families of the same quantity (given in brackets) will also be found from the context.

a	Atomic spacing, lattice parameter (a_0); microstructure-, composition- and blocking-related parameter in TRM expressions (a'); titanomagnetic composition coordinate; axis of ellipsoid.
A	Area of domain wall; constant related to exchange energy; empirical constant in $H_c(T)$; constant related to electron spin of ion ($\bar{A}$).
b	Lattice parameter; constant in law of approach to saturation; titanomagnetite composition coordinate; axis of ellipsoid.
B	Empirical constant in $H_c(T)$; constant related to electron spin of ion ($\bar{B}$).
c	Lattice parameter (c_0); titanomagnetite composition coordinate; axis of ellipsoid.
C	Curie constant.
D	Domain wall separation; vector for antisymmetric exchange (**D**); function of ionic concentration in spinel T_c model.
e	Electronic charge.
E	Potential energy (E_0, E_{anis}, E_{ex}, E_{self}, E_{ms}, E_w).
f_0	Frequency factor.
G	Structure dependent constant in domain wall energy.
h	Planck's constant ($\hbar$).
H	Magnetic field ($\mathbf{H}$, H_{mol}, H_{dip}, H_r, H_θ, H_d; applied fields—H_{ext}, H_s, H_0, H_1, H_2, H_3, H_p, H_{pB}, H_q; magnetization process parameters—H_c, H_{cr}, H_p, H^*).
i	Subscript; current in solenoid.
j	Subscript.

J Exchange energy constant (J_{ij}, J_{AB}, J_3).

k Wave number in fluctuating domain wall energy; Boltzmann's constant.

K Anisotropy energy constant (K_u, K_{u1}, K_{u2}, K_1, K_2, K_3, K_{si}, K_{dip}).

l Length; electron quantum number.

L Particle dimension (L_{crit}, $\bar{L}$); electron orbital angular momentum ($\mathbf{L}$); Langevin function ($\mathscr{L}$).

m Dipole moment ($\mathbf{m}$, m_{max}); index parameter in power law dependences; electronic mass; electron quantum number (m_l).

M Magnetization per unit volume ($\mathbf{M}$; material properties—M_s, M_A, M_B, M_1, M_2, M_3; induced magnetizations—M_{eq}, M_H, M_F, M_I; remanent magnetizations—M_r, M_{rs}, M_D, M_0); molecular weight.

n Number of domain walls in a particle; number of defects per unit volume ($\bar{n}$); saturation magnetization in β pfu (n_β); index parameter in power law dependences; electron quantum number; number of turns per unit length on a solenoid; particle size distribution function ($n(v)$).

N Demagnetizing factor ($N_\mathbf{a}$, $N_\mathbf{b}$. $N_\mathbf{c}$, N_θ); number of atoms/particles per unit volume (N^+, N^-, N_{TOT}); Avogadro's number (N_0).

p Volume concentration of dispersed grains; power law index parameter; electron population of $3d$ orbitals.

q Power law index parameter; electron population of $3d$ orbitals.

Q Konigsberger ratio (Q_T).

r particle radius (r_{crit}); polar coordinate; radius of electron orbit.

R Titanomagnetite non-stoichiometry parameter.

s Spin quantum number.

S Total electron spin quantum number of an atom (S_i, S_j, S_A, S_B); magnetic viscosity coefficient.

t Time (t_B).

T Temperature (T_0, T_1, T_2, T_a, T_R; critical/transition temperatures of materials—T_c, T_N, T_B, T_{Bi}, T_{trans}).

u Spinel structural parameter.

v Particle volume (v_i, v_B).

w Fraction of particles with volume v_i (w_i); function of angle between H and reference direction ($w(\phi)$).

W_r rotational hysteresis loss (W_{rp}).

x Displacement of domain wall (x_B); parameter in DRM model; composition coordinate of titanomagnetites (x', x_i); Fe–Mg cation distribution parameter; pyrrhotite composition coordinate (x_{sulphide}); composition coordinate in FeMgTi spinel.

y Haematite-ilmenite composition coordinate.

z Titanomaghemite composition parameter (z', z_B); number of nearest neighbours ($z_A, z_B, \bar{z}_A, \bar{z}_B$).

Z Atomic number.

α Direction cosine ($\alpha_1, \alpha_2, \alpha_3$); availability of Fe^{2+} spinel ions for oxidation; parameter in fluctuating domain wall energy; argument of Langevin function; volume parameter; angle in crystal structure.

β Bohr magneton; direction cosines($\beta_1, \beta_2, \beta_3$).

δ Domain wall thickness; cation distribution parameter.

Δ Crystal field splitting (Δ_0, Δ_T); function of ionic concentration in spinel T_c model.

ΔE_w Amplitude of fluctuating domain wall energy.

ε Angle in annular cone.

θ Angle between M and reference direction (θ_1, θ_2); angle of colatitude; asymptotic Curie temperature.

λ Magnetostriction coefficient ($\lambda_s, \lambda_{100}, \lambda_{111}, \bar{\lambda}$); molecular field coefficient (λ_w); spin-orbit coupling parameter.

μ_0 Permeability of free space.

σ Stress; magnetization per unit weight ($\sigma_0, \sigma_s, \sigma_H, \sigma_\perp, \sigma_\parallel$).

τ Relaxation/experimental time (τ^+, τ^-, τ_R).

ϕ Angle between H and reference direction (ϕ_1, ϕ_2, ϕ_3); angle of longitude.

χ Mass/volume susceptibility ($\chi_\perp, \chi_\parallel, \chi_{HF}, \chi_a, \chi_b$).

Contents

1 Introduction to the magnetism of rocks

1.1 Rocks as magnetic information stores

Technology imitates nature, often unwittingly. Many of the devices, processes, systems and materials invented by man are to be found in the animal and mineral world. Magnetic navigation in the animal world is an example recently added to the list which includes ranging using sound waves, servomechanics, information transfer by electrical signals, mechanical and optical devices (cantilevers, the ball-and-socket joint, lenses, the screw propellor), and composite materials. The student of the earth and life sciences can consider the Earth and cosmos not only as a laboratory in which experiments in physics, chemistry and biology are taking place, but also as a source of new technology. When V. Poulsen invented magnetic recording 83 years ago, the mineral world had anticipated him by several thousand million years. One wonders if Poulsen's experiments could have been inspired by the discovery of the magnetic recording properties of rocks only a few years before his patent was registered.

A magnetic record, and perhaps records carried by re-usable physical memory systems generally, are non-equilibrium states in that the recorded conditions have passed away, and the system does not respond to, and is not in equilibrium with, the current conditions. Otherwise the state of the system would not constitute a 'memory'. The memory system must be perturbed to allow it to respond to the external conditions and the perturbation suppressed to maintain the record. No memory is truly permanent. An electronic memory system is perturbed by allowing voltage pulses to appear at its inputs, and an accurate record preserved by reducing the probability of occurrence of noise-generated or other input pulses at other times. Information is written into the elements of the obsolescent digital magnetic core store by applying a large enough magnetic field that the resistance of a core to changing its magnetization is overcome. The resistance to magnetization must be high enough to reduce the probability of any element changing its magnetic state during the writing operation for another core. The memory may be degraded later due to the effects of noise—either magnetic fields or thermal agitation in the material. In the case of 'moving surface' magnetic memories, the perturbation is the exposure of the surface to the recording head. Again the later fidelity of the record depends on the levels of perturbing noise and thermal agitation. The lower the temperature, the lower the levels of thermal noise in electrical or magnetic systems.

This pattern, the application of a perturbation to imprint the information with the subsequent fidelity of the record depending on how effectively the perturbation is later suppressed and on the level of internal noise, also applies to the mechanisms by which the magnetic minerals in rocks function in nature as memory systems. Taking first the case of a sedimentary rock being formed by deposition of mineral grains in a lake or sea, the perturbation is the mechanical freedom which the magnetic mineral grains enjoy before the sediment is compacted. The magnetic mineral grains will probably possess a magnetic moment, and this results in a torque tending to turn each grain to align its moment in the field direction while the perturbation operates. Other, mechanical, effects tend to randomize the grains. Some equilibrium state is achieved. After compaction, i.e. removal of the perturbation, the grains can no longer rotate in response to a torque, and the resultant magnetization of an assemblage of the grains will be orientated in the field direction. A remanent magnetization thus acquired is termed a *depositional remanent magnetization* (DRM). The record is preserved provided the mechanical integrity of the sediment is maintained and the effect of subsequent thermal noise is small, the latter depending on temperature and how difficult it is to change the magnetic state of the individual grains.

In the case of igneous or metamorphosed rocks, the perturbation is the thermal noise itself, associated with the elevated temperature from which the rocks cool. The thermal vibrations allow the magnetization vectors of the assembled magnetic grains to come into equilibrium with the weak ambient geomagnetic field. On cooling to lower temperatures the perturbation may be effectively removed, and the system has a *thermoremanent magnetization* (TRM). The imprint is preserved so long as the excursion to elevated temperature is not repeated and depending on the effect of thermal noise at room temperature.

A *chemical remanent magnetization* (CRM) is acquired when magnetic grains nucleate and grow in a field. Here the perturbation is the small grain volume. This results in the magnetic state of the grain being easily changed, and it is thus susceptible to thermal noise. Again a statistical alignment of grain vectors, corresponding to equilibrium with the ambient field, results in an imprint once the perturbation is removed, i.e. the grain grows large enough that thermal noise becomes ineffective. The moment is preserved as long as the grains are not reduced in size or some other perturbation is applied.

A system in which the grains have low resistance to magnetization change, so that thermal vibrations significantly perturb the assemblage continuously even at room temperature, cannot operate as a 'permanent' memory. In the context of rocks, the state of such a system inevitably corresponds to the present-day geomagnetic field and the remanent state is a *viscous remanent magnetization* (VRM).

Other perturbations which may have importance in the rock magnetic memory systems are *pressure* (giving PRM) and *shock* (SRM).

1.2 The recorded information

The direction of the field experienced by the rock is recorded at the time of relaxation of the perturbations. A rock experiencing several perturbations may carry several records depending on the extent to which the earlier records are erased by the later. When contemporaneous ancient geomagnetic field directions from different parts of the Earth's crust are compared, the conclusion is that the predominant dipolar symmetry of the geomagnetic field has been maintained throughout geological time, but that the polarity of the field frequently changes. Once the dipolar field is taken as a datum, ancient field directions can be interpreted in terms of ancient latitudes and orientations with respect to the meridian of the continental masses.

The direction of magnetization of large samples of the Earth's crust is contained in the field measured in large-scale magnetic surveys of the Earth. Although the determination and interpretation of these recorded directions is much more difficult than for the collected palaeomagnetic samples, the gross directions—normal or reverse—seem to be determinable, at least for the younger crust of the ocean basins. The relative dispositions of sections of the submarine crust believed to carry contemporaneous magnetizations can be inferred.

The remanence of rocks also records the time of the perturbation, if the time dependence of the field direction is known. This provides a method of age determination and information on the kinematics of the tectonic processes.

The intensity of the remanence is related to the intensity of the field and to the nature of the perturbation. Information about both can, in principle, be extracted from the natural remanent magnetization (NRM) of rocks, or artefacts such as the baked clay of hearths, bricks and pottery. The latter also contain directional information if undisturbed since their last heating in antiquity.

The study of the NRM of rocks (palaeomagnetism and palaeointensity studies) contributes to geomagnetism through the derived history of the direction and intensity of the field, and to tectonics through the inferred history of the disposition of the recording medium. The study of the NRM of artefacts (archaeomagnetism) contributes to our picture of the geomagnetic field in recent times and may produce information of archaeological value.

1.3 Rock and mineral magnetism

A major part of the research field of 'rock and mineral magnetism' is concerned with magnetic minerals in rocks as a recording medium, and with the recording mechanisms by which a faithful record can be maintained over geological time. Quantitative models for the mechanisms are sought which are appropriate to the perturbations rocks undergo during their formation and subsequent history. The models are subjected to experimental tests. As always, models are simple compared to the reality, whether it is a real rock or an artificial one prepared in the laboratory. In addition to the properties of assemblages of magnetic grains as acquirers of remanence in weak fields, rock and mineral magnetism is also concerned more generally with the magnetization process parameters of minerals, and interpreting such parameters in terms of the intrinsic or material properties of the minerals and their microstructure. The intrinsic properties are a prerequisite in identifying magnetic minerals in rocks by magnetic means, and one of the magnetization process parameters—the initial susceptibility—has direct relevance in interpreting the results of magnetic surveys.

If models for the weak-field remanence mechanisms provide the physical basis of palaeomagnetism and its applications, the chemical basis—knowledge of the petrological processes which magnetic minerals undergo during the formation of rocks and at later times, and which ensure that they become receptive media for the palaeomagnetic signal—is equally important. 'Magneto-mineralogy' lies within the ambit of rock and mineral magnetism. The medium itself, therefore, is also to be studied. This involves the determination of the composition, concentration and microstructure (principally grain size) of the magnetic mineral fraction in rocks. It may then be possible to declare on the performance of a particular rock, or rock type, to assign the components of a multicomponent remanence to different perturbations and mineral fractions, and comment on their likely stability.

Just as rock and mineral magnetism draws on the techniques and experience of solid state magnetism and experimental petrology, so the developed techniques of rock magnetism find applications in fields other than geophysics and geology. One example is the growing use of magnetic methods in 'environmental science'. Such studies involve tracing the movement of magnetic minerals between the atmosphere, hydrosphere and lithosphere. Man plays a role here in burning fossil fuels and clearing land for agricultural use. More localized environmental problems with a magnetic angle are found in the industrial health and public health areas, when magnetic material enters the lungs of welders or the environs of a steel works are polluted by its effluent. Workers in the newly emerging research field of 'biomagnetism' have also turned to rock magnetism for the means of investigating the small magnetic particles which form part of the nervous systems of many animals.

1.4 Magnetism—some preliminaries

Magnetic fields have two sources—the flow of electrical current and the 'leakage' field from a magnetic body. The former we believe to be the origin

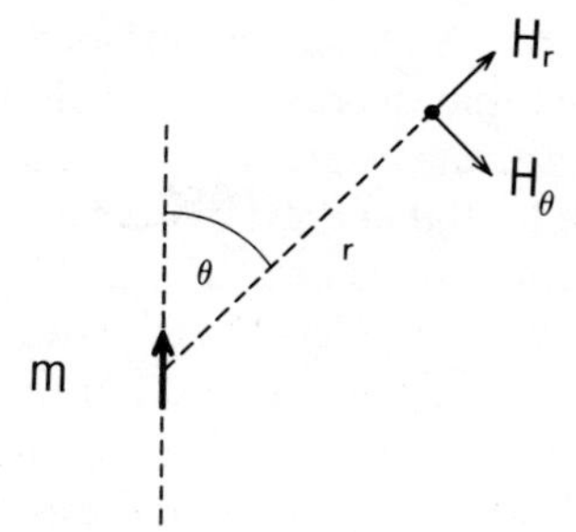

Figure 1.1 The magnetic field due to the model dipole at the point (r, θ) expressed in terms of the radial and tangential components H_r and H_θ.

of the Earth's magnetic field, and the latter is the source of magnetic 'anomalies' in the field measured near the Earth's surface. Of course, we see magnetism and magnetic materials at work almost everywhere in the home, car and in places of work and leisure. The magnetic field H along the axis of a long solenoid which has n turns per unit length and carries a current i amps (A) is $H = ni$. The units of magnetic field are A/m. The field due to a magnetized body is in general complicated, as is the spatial variation of the field in the vicinity of any real current-carrying conductor. For an observer at a distance which is large compared to the dimensions of a body or current-carrying system of simple form, the field can be represented as the field of the model dipole. The general shape of the dipolar field is familiar from elementary physics experiments. The spatial variation of the field is best expressed in the planar polar (r, θ) coordinates suited to the uniaxial symmetry of the dipole (Fig. 1.1). The field at point (r, θ) is the vector sum of the two components $H_r = 2m \cos\theta/4\pi r^3$ and $H_\theta = m \sin\theta/4\pi r^3$. The size and orientation of the resultant can readily be worked out. The dimensions of m, the dipole moment, are Am^2. A magnetized body has magnetization M where this is the equivalent dipole moment of the body divided by its volume. M has dimensions A/m.

The magnetization vector of a magnetized body experiences a torque when placed in a uniform magnetic field. When the magnetization vector is fixed within the body, the torque is transferred to the body itself, whence the compass needle. The torque is concomitant with an angular dependence of potential energy of the magnetic vector in the field, i.e. work must be done to rotate the vector out of parallel alignment with the field. The potential energy is given by the scalar product of the two vectors and equals $-\mu_0\mathbf{m}.\mathbf{H}$ joules or $-\mu_0\mathbf{M}.\mathbf{H}$ joules/m^3 where μ_0 is a constant of nature ($= 4\pi \times 10^{-7}$ henry/m in MKS). More simply, this is $-\mu_0 mH\cos\theta$ or $-\mu_0 MH\cos\theta$. This expression can be regarded as being empirical, and it is evident that the form of the expression, $-\cos\theta$, is in accord with our experience of magnets and fields (Fig. 3.7).

Apart from exerting a torque on a magnetized body, the magnetic field may also change the internal state of the body or a system of bodies. This is directly due to the potential energy of a dipole in a field; the change in internal state—the magnetization process—is the body, or the system, in the act of reducing the potential energy. For small fields, the magnetization change is approximately linearly related to the field, $\mathbf{M} = \chi\mathbf{H}$. χ, the magnetic volume susceptibility, is dimensionless. In general, χ is a function of H and may not be single valued.

Magnetism is currently undergoing a transition from the CGSEMU system to MKS units. Although at one time the Kennelly magnetism MKS system was fashionable—Chikazumi's excellent textbook, for example, is written in Kennelly MKS—the Sommerfeld system now seems to be being adopted. The two systems are equivalent in practical terms, differing only in μ_0 being located sometimes on the numerator of expressions in one system and on the denominator in the other; sometimes present in one and absent in the other. One transitional factor in favour of the Sommerfeld system is that the Sommerfeld dipole moment per unit volume is related to the CGS magnetization by a factor 10^3. As MKS density is related to CGS density also by a factor of 10^3, the magnetic dipole moment per unit mass (Am^2/kg) has the same numerical value in both the Sommerfeld and CGS systems. This parameter is the only one for which the practitioner can carry forward a 'feeling' for the magnitude of quantities. For magnetization per unit volume, dipole moment, mass and volume susceptibilities, coercive force and magnetizing fields, the magneticist must start again from scratch. The numerical sizes of quantities in the Sommerfeld system seem to be easily digestible. The remanence per unit volume of a strongly magnetic rock (which is still, of course, weak compared to a man-made permanent magnet) is of the order of 1 A/m. The volume susceptibility of such a rock might be of the order of 10^{-2}. This can be compared to the volume susceptibility (~ 1) of a more or less equidimensional 'soft' and strongly magnetic body. The mass susceptibility of a paramagnetic salt is of the order of 10^{-6} m^3/kg and a diamagnetic material (such as water) $\sim -10^{-8}$ m^3/kg. The intense magnetic field of a laboratory electromagnet is about 10^6 A/m. 10^5 A/m can be obtained from a solenoid without too much difficulty. 10^5 A/m is about the highest coercive force observed in minerals. The geomagnetic field is of the order of 10 A/m.

2 Magnetic minerals in rocks

2.1 The magnetic mineral systems

The primary source material from which the Earth's crust is constructed is the volcanic and plutonic rock of the spreading centres of the ocean basins. The primary magnetic minerals of the crust are the minerals of such rocks. Just as the rocks are altered in the submarine environment, reworked at the subduction zones, weathered and eroded as continental igneous rocks, deposited as sediments, metamorphosed, eroded and so on, so do the magnetic minerals undergo a long and complicated history over geological time. The magnetic minerals are compounds of iron, this being the most abundant of the transition elements. The iron of the primary rocks is distributed between both magnetic and non-magnetic phases. In the course of geological time, the iron of the primary magnetic phases may come to be contained in non-magnetic phases and vice versa. These transformations may occur many times over in the course of alteration, solution, precipitation and metamorphism.

Many of the magnetic species, the oxide minerals, may be represented, at least approximately, on the FeO–TiO_2–Fe_2O_3 ternary diagram (Fig. 2.1). In fact iron and titanium may not be the only metal species in oxide minerals and a multidimensional 'composition space' is needed to represent the composition of minerals or mineral mixtures, or to track the path taken by a mineral during its evolution. In transformations to non-magnetic phases the mineral may disappear from this diagram. In transformations producing a magnetic oxide the mineral may appear on the diagram having come from a point outside. The diagram essentially displays the relative proportions of Fe^{2+}, Fe^{3+} and Ti^{4+}. The lower right-hand corner is therefore taken as $\frac{1}{2}Fe_2^{3+}O_3^{2-}$ or $Fe^{3+}O_{3/2}^{2-}$ and all points in the triangle represent a cation mixture or solution adding up to one cation. The primary magnetic minerals (immediately post-crystallization) are the titanomagnetites with approximate composition indicated by the point TM60 in Fig. 2.1. During geological history the composition of the magnetic phase will generally move to the right (corresponding to oxidation) and downwards, haematite being the ultimate stable magnetic phase. Some of the processes which may take place are indicated in the figure. The titanomagnetite may oxidize at high temperature ('deuteric oxidation') producing an intergrowth of which the magnetic component is a titanomagnetite richer in iron than TM60. Low temperature oxidation ('maghemitization') produces a titanomaghemite, a structurally

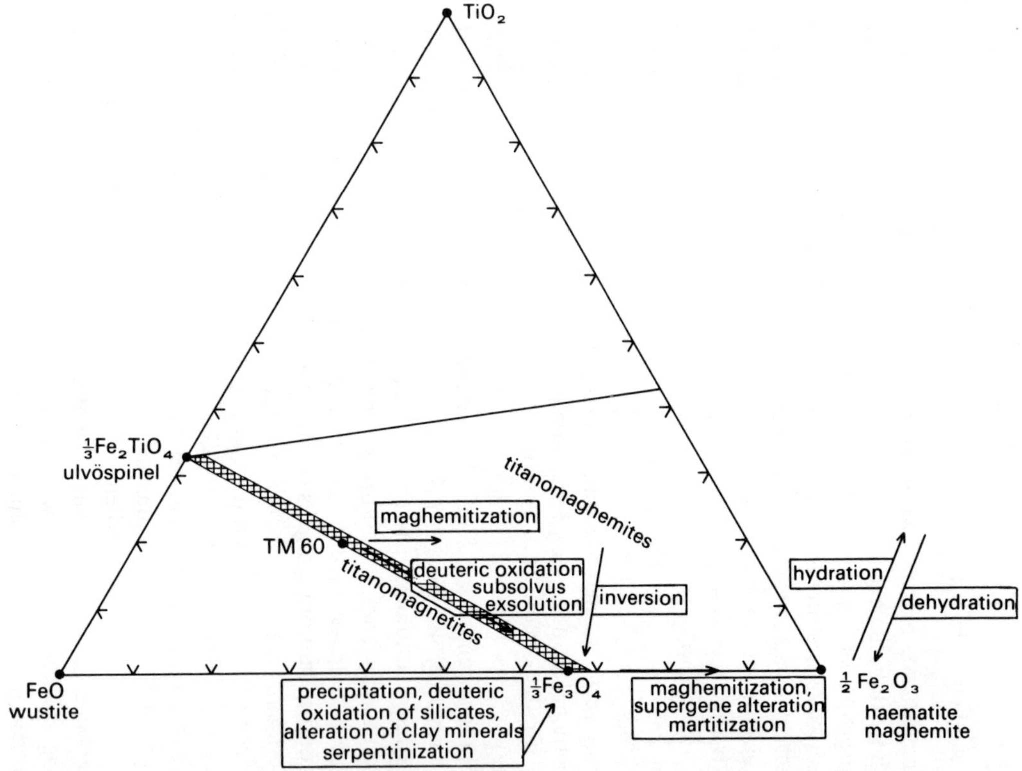

Figure 2.1 The FeO–TiO_2–Fe_2O_3 ternary diagram of rock magnetism. Points in the triangle correspond to oxide compounds or mixtures of compounds containing Fe^{2+}, Fe^{3+} and Ti^{4+}. The point TM60 represents the idealized titanomagnetite with composition $Fe_{2.4}Ti_{0.6}O_4$ and on the diagram are indicated some of the paths which may be followed by the magnetic mineral fraction of crustal rocks in the course of geological history.

metastable mineral which may 'invert' to produce magnetite or haematite. A slow-cooled titanomagnetite may undergo subsolvus exsolution producing an intergrowth of titanomagnetites one richer in iron and one poorer in iron than TM60. Magnetite may be produced by the deuteric alteration or the serpentinization of non-magnetic iron-containing silicates, by alteration of non-magnetic clay minerals, or by precipitation from iron-bearing solutions. Magnetite may in turn be maghemitized to produce a maghemite-magnetite solid solution which may invert to a haematite-magnetite intergrowth, or direct oxidation of magnetite may produce such an intergrowth. The leaching of iron from minerals produces iron-bearing solutions from which weakly magnetic or non-magnetic iron oxyhydroxides may precipitate. These may in turn dehydrate to produce haematite or maghemite. Many possibilities are presented by nature during geological time.

Before considering some of these processes in more detail or in their petrological context we shall first deal with the crystal chemistry of the individual mineral systems.

2.1.1 *The titanomagnetites*

The formula unit of the ideal titanomagnetite is $Fe_{3-x}Ti_xO_4 (0 < x < 1)$ which is a member of the solid solution between magnetite, Fe_3O_4, and ulvöspinel, Fe_2TiO_4. The $(3 - x)$ iron ions are divided into $(1 + x)$ Fe^{2+} and $(2 - 2x)$ Fe^{3+} per formula unit (pfu). The real titanomagnetite, whether occurring in nature or as a synthetic material, deviates from the ideal in containing other species of cation than iron and titanium and in being non-stoichiometric (i.e. a cation/anion ratio not equal to 3/4). The real material can be regarded as having been obtained from the ideal by substitution and oxidation/reduction. Al^{3+}, Mg^{2+} and Mn^{2+} (in concentrations of the order of 0.1–0.5 pfu in titanomagnetites in basalts) can be considered as having replaced Fe^{3+} and Fe^{2+}. The formula unit of a substituted stoichiometric titanomagnetite may be written $Fe_{3-x-a-b-c}Al_aMg_bMn_cTi_xO_4$ with $(1 + x - b - c)$ Fe^{2+} and $(2 - 2x - a)$ Fe^{3+} pfu. Oxidation of the titanomagnetite will produce a cation/anion ratio less than 3/4, a corresponding change in the Fe^{3+}/Fe^{2+} ratio maintaining charge neutrality in the crystal. If the cation/anion ratio has fallen to a fraction R of the ideal value, i.e. equals $3R/4$, the formula unit may be written $Fe_{(3-x-a-b-c)R}Al_{aR}Mg_{bR} \cdot Mn_{cR}Ti_{xR}O_4$. There are now $(1 + x - b - c)R - 8(1 - R)Fe^{2+}$ and $(2 - 2x - a)R + 8(1 - R)Fe^{3+}$ pfu. It can be seen that R can take values between 1 (stoichiometry) and $8/(9 + x - b - c)$ when no Fe^{2+} remains. Although R is a useful and convenient parameter for describing deviations from stoichiometry in these compounds, a parameter which describes the degree of oxidation more obviously is the z composition parameter which will be discussed in the context of titanomaghemites (section 2.1.2).

The titanomagnetites crystallize in the spinel structure. This consists of a

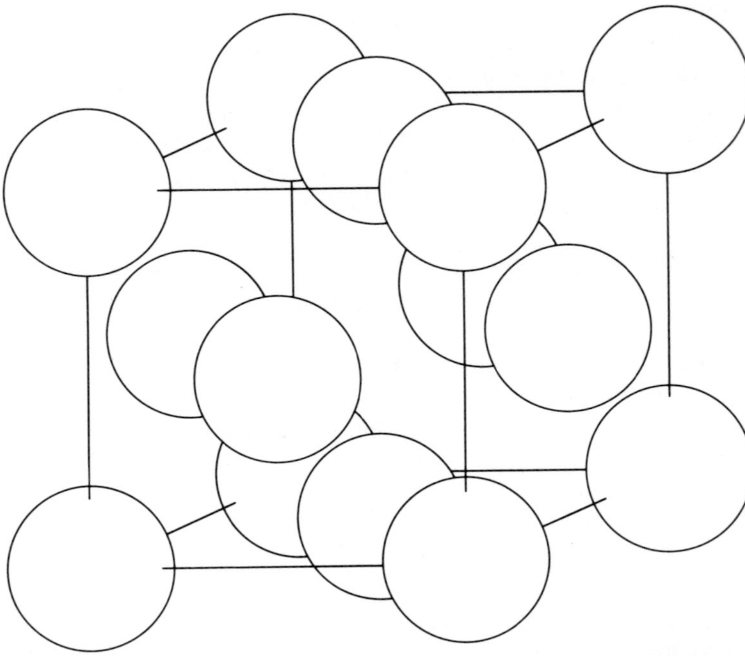

Figure 2.2 The unit cell of the face-centred cubic system. The close packed layers may be seen forming planes normal to the body diagonal, ⟨111⟩ direction. The cell contains the equivalent of 4 atoms or ions with 8 interstices of tetrahedral and the equivalent of 4 interstices of octahedral symmetry.

framework of cubic close-packed anion layers, the layers forming (111) planes of a face-centred cubic (fcc) lattice (Fig. 2.2). The framework encloses interstices of two symmetries having four and six nearest neighbours. Eight sites of four-fold symmetry (tetrahedral sites) are contained in the fcc unit cell, one at the centre of each octant cube. Octahedral sites of six-fold symmetry lie at the centre of the cell and halfway along each cube edge. In simple ideal monoxide compounds, all octahedral sites are occupied by divalent cations. In spinel oxides containing cations with a range of valencies (but with an average valency of 8/3) only 1/8 of the tetrahedral interstices and 1/2 of the octahedral interstices are occupied. In this special arrangement each anion has one tetrahedral site cation and three octahedral site cations as nearest neighbours. The nearest neighbours of each of the three kinds of site are shown in Fig. 2.3. The next-nearest (octahedral site cation) neighbours of octahedral sites are also shown for the central site in the figure. An element of trigonal symmetry can be seen, which receives further contributions from the puckering of the anion layers which invariably occurs to some extent. The anions are displaced towards the centres of the octant cubes shown in the figure. The axis of this trigonal symmetry for the central octahedral site is shown by the dashed line normal to the (111) planes. The positions of the

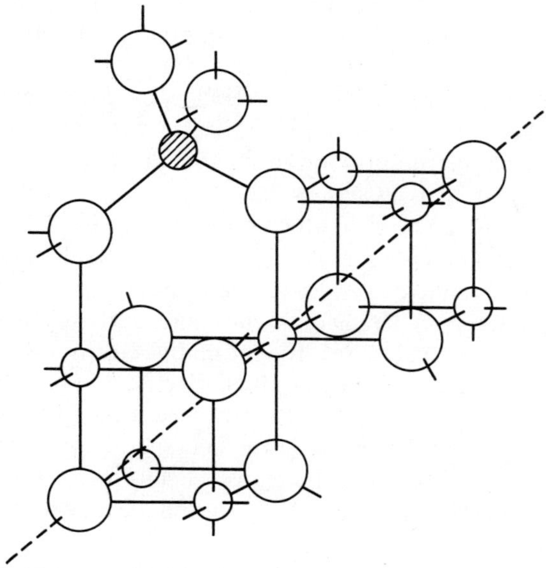

Figure 2.3 A part of the spinel structure showing the environs of the two types of cation site and the anion sites. Large open circles represent anions, small open circles cations in octahedral sites, and the small hatched circle a cation in tetrahedral symmetry. In practice the cubic anion framework is distorted, the anion tetrahedra surrounding tetrahedral sites expanding and the anion tetrahedra interlocked with octahedral sites contracting.

oxygens are described by the *u* parameter (the distance of an oxygen from a reference cation (100) plane) which may be taken as 0.375 units for an ideal array. Measured values of *u* range from 0.370 to 0.390. A relatively large unit cell is needed to contain an average cation population, eight times the volume of the fcc unit cell of Fig. 2.2. The unit cell contains eight formula units $A|B|_2O_4$ where A refers to tetrahedral site and B to octahedral site cations. In non-stoichiometric oxidized spinels some of the cation sites are empty or may be considered as being occupied by vacancies of zero valency, □.

The distribution of cations between the tetrahedral and octahedral sites of the spinel structure is of some interest and importance. In the case of the titanomagnetites, the cation distributions for the end members of the solid solution are written $Fe^{3+}|Fe^{3+}Fe^{2+}|O_4$ and $Fe^{2+}|Fe^{2+}Ti^{4+}|O_4$, the octahedral site population shown in parenthesis. The minimization of a number of competing contributions to the total lattice energy determines the cation distribution in spinels. The electrostatic energy, the (negative) work done to assemble the array of charged ions, is reduced by having low charge ions (e.g. Fe^{2+}) on tetrahedral sites. This produces a larger electric field at anion sites and the consequent polarization of the anions is a favourable effect. Individual site preferences also exist because of the form of the transition element electronic orbitals (see section 3.1., Chapter 3) and their interaction with the electrostatic field produced by the nearest neighbours in tetrahedral or octahedral sites. Although this may explain the preference of Cr^{3+} for octahedral symmetry, zero or weak preference is expected for Fe^{3+} and Fe^{2+}. Treatment of spinels as non-ionic compounds, in which molecular orbitals are formed, is necessary to theoretically describe the tetrahedral site preference of Fe^{3+} which apparently dictates the cation distribution in Fe_3O_4. The cation distribution in the titanomagnetite solid solution seems variable (see section 7.1.1.1, Chapter 7).

In magnetite Fe^{3+} and Fe^{2+}, which differ only in Fe^{2+} having one extra electron, occur on the octahedral sublattice in equivalent sites. Adequate thermal energy is available at room temperature to effect the transfer of the extra electron between the Fe ions, and the octahedral sites can be considered as being occupied by averaged Fe ions. The mobile electrons hopping between octahedral sites produce a relatively high electrical conductivity and may contribute to the progress of maghemitization (section 2.1.2). At about 120 K some kind of ordering takes place in the octahedral sublattice (the Verwey transition) and the crystal structure transforms from cubic to a lower symmetry. Octahedrally sited Fe^{2+} and Fe^{3+} then make individual contributions to physical properties.

The cation-oxygen bond length is found to be approximately constant in a given coordination in a wide range of compounds. An alternative expression

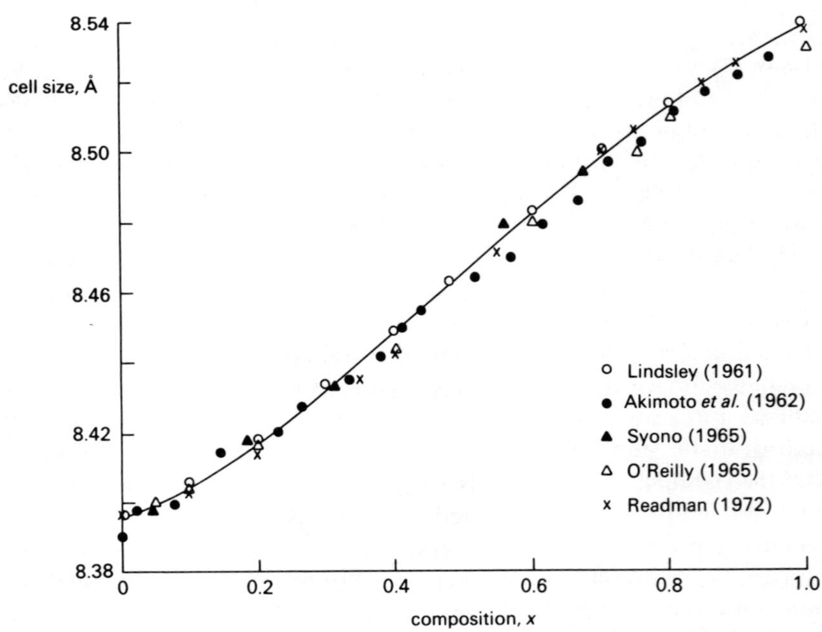

Figure 2.4 The unit cell parameter (Å) of synthetic titanomagnetites as determined by a number of investigators. Variations are probably due to variations in high temperature non-stoichiometry, lower values indicating greater degrees of non-stoichiometry.

for this observation is found in the assignment of definite values to ionic radii and in the 'rigid sphere' model of solids. In the case of spinels the unit cell expands or contracts to accommodate larger or smaller cations. The titanomagnetite unit cell increases from about 8.395 Å for magnetite to 8.535 Å for ulvöspinel (see Fig. 2.4). The cell edge of $Fe_{2.4}Ti_{0.6}O_4$ ($x = 0.6$, TM60) is 8.485 Å giving an (ideal) X-ray density of 4.84×10^3 kgm/m^3.

2.1.2 *The titanomaghemites*

No definite boundary can be said to exist between titanomaghemites and non-stoichiometric titanomagnetites. However, distinction can be drawn in that titanomagnetite crystallizing from the melt and cooling from high temperature can sustain only a limited degree of non-stoichiometry before breaking up into an intergrowth of phases (deuteric oxidiation, see section 2.2). At low temperature, oxidative alteration (maghemitization) can produce high or the maximum possible degrees of non-stoichiometry. Titanomagnetite/titanomaghemite with either high or low temperature non-stoichiometry can be described by the same formula. The ideal titanomagnetite is $Fe_{3-x}Ti_xO_4$ $(0 < x < 1)$ and the ideal non-stoichiometric titanomagnetite/titanomaghemite $Fe_{(3-x)R}Ti_{xR}\square_{3(1-R)}O_4$ where now x may not have the same limits (see below) and R may take values between 1 (stoichiometry) and $8/(9 + x)$. $\square$ are vacancies occurring in the cation sites of the structure. As the titanomaghemites are essentially produced by oxidation of titanomagnetites, their crystal chemistry depends on the mechanism by which the metal: oxygen ratio is made to change during the oxidation. Studies of titanomaghemite in submarine basalts (section 2.2) show that the ratio may be changed by one or both of the mechanisms of addition of oxygen or removal of metals. Let us consider how the composition changes under the two mechanisms.

(i) *Addition of oxygen.* In this mechanism, oxygen atoms are adsorbed at the surface of the titanomagnetite grain. The locally mobile electrons in the process of transferring between Fe^{2+} and Fe^{3+} at the surface are available to reduce the oxygen atoms to ions which become chemically bound to, and form an extension of, the crystal lattice. The surface will be richer in Fe^{3+} and vacancies and poorer in Fe^{2+} and Ti^{4+} than the interior of the grain and the concentration gradients will drive systematic ionic and electronic diffusion within the grain to reduce such gradients. Eventually a near-homogeneous titanomaghemite grain will be produced. The $Fe^{3+}:Fe^{2+}$ ratio increases due to the loss of electrons to the oxygen; the metal: oxygen ratio decreases leading to vacant cation lattice sites and the Fe:Ti ratio remains constant. Each Fe^{2+} needs $(\frac{1}{2})$O to effect the change from FeO to $FeO_{3/2}$. A formula unit of titanomagnetite contains $(1 + x)Fe^{2+}$ and therefore requires $(\frac{1}{2})(1 + x)$ O for complete oxidation. We can introduce an oxidation parameter z which describes how far this reaction has gone towards completion. The addition of $(z/2)(1+x)$ O will have oxidized a fraction z of the initial Fe^{2+} producing an additional $z(1+x)$ Fe^{3+}. For every formula unit (4 oxygens) in the original grain, there are now $\{4+(z/2)(1+x)\}/4(=1/R)$ formula units.

The reaction may be summarized:

at the surface $Fe^{2+} \leftrightarrow Fe^{3+} + 1e$

$$O + 2e \rightarrow O^{2-}$$

for the system as a whole

$$Fe_{3-x}Ti_xO_4 + (z/2)(1+x)O \rightarrow (1/R)Fe_{(3-x)R}Ti_{xR}\square_{3(1-R)}O_4 \qquad (2.1)$$

where use of the R parameter conveniently makes what would otherwise be a cumbersome formula relatively simple. The $(3-x)R$ Fe pfu is made up of $(1-z)(1+x)R$ Fe^{2+} pfu (the residual Fe^{2+} spread throughout the extended lattice) and $\{2-2x+z(1+x)\}R$ Fe^{3+} pfu.

The path taken by the evolving mineral may be traced on the ternary diagram (Fig. 2.5). The composition moves to the right along the horizontal lines. These lines correspond to a constant composition coordinate x and variable coordinate z increasing from zero to unity. Lines of constant z and variable x are also shown.

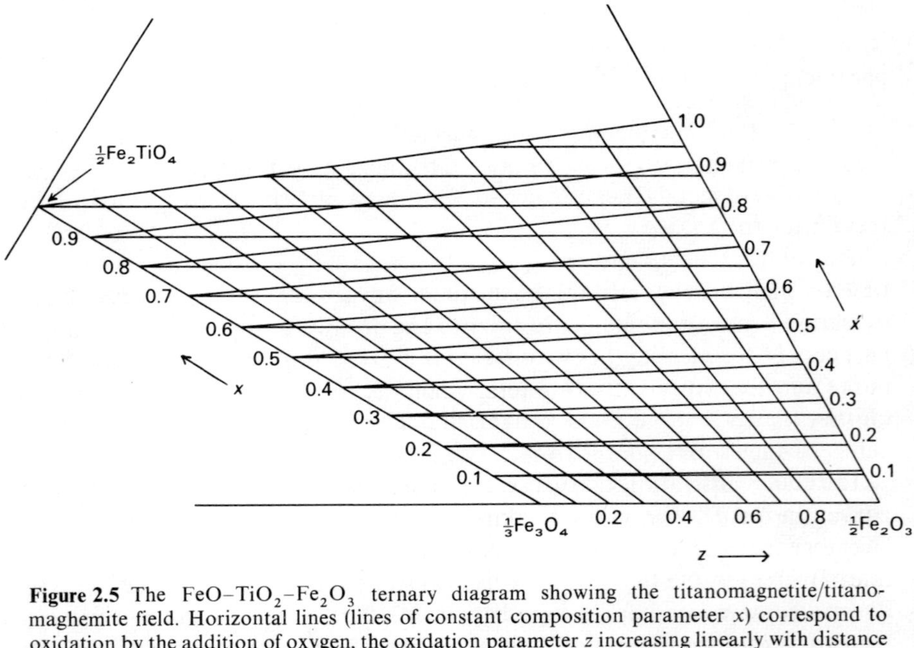

Figure 2.5 The FeO–TiO_2–Fe_2O_3 ternary diagram showing the titanomagnetite/titanomaghemite field. Horizontal lines (lines of constant composition parameter x) correspond to oxidation by the addition of oxygen, the oxidation parameter z increasing linearly with distance along a line of constant x. The inclined lines indicate the path followed on oxidation by removal of iron. Points in the main part of the titanomaghemite quadrilateral may be reached by either mechanism or a combination of the mechanisms.

(ii) *Removal of metals.* For this mechanism oxygen atoms are not present at the surface and the addition of oxygen mechanism is therefore inhibited as locally mobile electrons at the surface, hopping between Fe ions, do not encounter any adsorbed species in a reduced state. It is envisaged that the locally mobile electrons may momentarily accumulate in sufficient numbers on an individual Fe cation site to produce the atomic form. When this occurs at the surface, such atoms being no longer chemically bound may be leached away or react with the surrounding medium before reverting to the ionic state. Diffusion within the grain again is required to bring about homogeneity. The $Fe^{3+}:Fe^{2+}$ ratio increases due to the loss of electrons with the escaping atoms; the metal: oxygen ratio falls leading to vacant lattice sites; the Fe:Ti ratio falls and the number of spinel units in the grain remains constant.

The reaction may be summarized:

at the surface $Fe^{2+} \leftrightarrow Fe^{3+} + 1e$

$$Fe^{3+} + 3e \leftrightarrow Fe$$

$$Fe^{2+} + 2e \leftrightarrow Fe$$

the latter two reactions being equivalent as, in either case, the production of one Fe atom requires the loss of $3Fe^{2+}$ and the gain of $2\,Fe^{3+}$ (and, of course, charge is conserved). Oxidation may also take place by removal of Ti but this may require the simultaneous presence of Ti^{3+} and Ti^{4+} and an electron hopping process. The maximum loss of iron for a titanomagnetite with $(1+x)\,Fe^{2+}$ pfu is $(1/3)(1+x)$ atoms. We can introduce an oxidation parameter z' which describes how far this reaction has gone towards completion. Thus for the system as a whole

$$Fe_{3-x}Ti_xO_4 \rightarrow Fe_{3-x-(z'/3)(1+x)}Ti_x\square_{(z'/3)(1+x)}O_4 + Fe_{(z'/3)(1+x)} \quad (2.2)$$

there now being $1 + x - z'(1+x)\,Fe^{2+}$ and $2 - 2x + (2z'/3)(1+x)\,Fe^{3+}$ pfu. Again we have movement to the right on the ternary diagram, however this time on inclined lines (Fig. 2.5), the composition coordinate x increasing as oxidation proceeds. Points in the main body of the titanomaghemite quadrilateral may be reached by either mechanism or a combination of both. Points in the upper part of the quadrilateral (above $x = 1, 0 < z < 1$) can be reached only by the removal of iron mechanism. Such points represent new minerals yet to be found in nature.

The composition coordinates (x, z) can usefully be extended into the upper part of the quadrilateral, x now taking values greater than unity (up to 1.285). The lines of constant x are still equispaced and horizontal and z varies linearly along a line of constant x from zero on the titanomagnetite line (or its extension) to unity at the right hand side of the ternary diagram. The x in the titanomaghemite formula unit of equation (2.2) is not the x coordinate of the titanomaghemite composition, but that of the starting titanomagnetite. Confusion is avoided by replacing the x of the titanomaghemite of equation

(2.2) by the parameter x' which is invariant during oxidation by removal of iron, just as x remains constant during oxidation by addition of oxygen. Thus

$$Fe_{3-x}Ti_xO_4 \rightarrow Fe_{3-x'-(z'/3)(1+x')}Ti_{x'}\square_{(z'/3)(1+x')}O_4 + Fe_{(z'/3)(1+x')} \quad (2.3)$$

The inclined lines of Fig. 2.5 are lines of constant x' and points in the quadrilateral may be defined by either coordinates (x, z) or (x', z'). z', the fraction of reaction for removal of iron, is not the same as z, the fraction of reaction for addition of oxygen even in the case of the magnetite-maghemite line (the base of quadrilateral) for which the two mechanisms follow the same path. The two parameters are equal only at zero and unity. z' does not plot linearly along lines of constant x'. The (x, z) coordinates are to be preferred for display of mineral compositions on the diagram. However, (x', z') are useful working coordinates for experimental studies or theoretical consideration of the removal of iron mechanism. The relations between (x, z) and (x', z') are easily determined by equating the titanomaghemites of equations (2.1) and (2.3). We obtain $z = 8z'/(9 - z')$ and $x = 9x'/\{9 - z'(1 + x')\}$. We also have $x' = xR$ and $x = x'(8 + z)/(8 - zx')$.

(iii) *Structural properties of titanomaghemites.* The distribution of cations and vacancies between the sublattices of the spinel titanomaghemites probably depends on both the cation distribution in the initial titanomagnetite and the maghemitization mechanism. The physical properties, insofar as they depend on cation distribution, will not therefore be unique for a given titanomaghemite composition (see section 7.2.1) except possibly for the special case of the magnetite–maghemite solid solution. It seems probable that the cation distribution in maghemite ($\gamma - Fe_2O_3$ or $Fe_{8/3}\square_{1/3}O_4$) is $Fe^{3+}|Fe^{3+}_{5/3}\square_{1/3}|O_4$ and that compositions intermediate between $\gamma - Fe_2O_3$ and Fe_3O_4 have only Fe^{3+} in tetrahedral sites and vacancies only in octahedral sites. In the case of the titanomaghemites the presence of vacancies (and Ti^{4+}) only in octahedral sites is often adopted as a working hypothesis.

The spinel unit cell contracts following maghemitization, consistent with the increased concentration of Fe^{3+} and $\square$. Some variation is found in the cell edges determined for synthetic titanomaghemites. The results of one investigation are shown in Fig. 2.6 in the form of contours of constant cell edge. Extra reflections are observed on X-ray powder pictures of maghemite and some titanomaghemites which may be ascribed to an ordering of cations and vacancies on octahedral sites producing a tetragonal superstructure.

The open structure of the titanomaghemites is metastable and collapses to a more stable configuration on application of hydrostatic pressure or on heating to moderate temperatures. In the simplest case, maghemite transforms ("inverts") to haematite under 1.5×10^7 pascals (150 bar) at 0°C or at temperatures of about 350°C at atmospheric pressure. The inversion is effected by a displacement of the oxygen layers from the cubic close packing of the spinel structure to the hexagonal close packing of the corundum

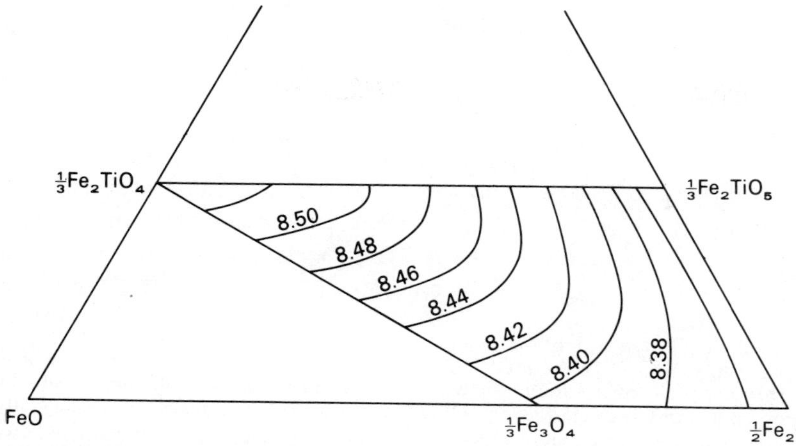

Figure 2.6 The unit cell parameter (Å) for titanomaghemites synthesized by the addition of oxygen mechanism (O'Reilly and Readman, 1971). Cell edge is not very sensitive to cation distribution and it is probable that no great differences would result from maghemitization by the alternative mechanisms.

structure (section 2.1.3) and a reorganization of the Fe^{3+} cations in the mixed trigonal and kagome layers of the spinel structure. Solid solutions of magnetite and maghemite invert to intergrowths of magnetite and haematite. The titanomaghemites invert to a more complicated mixture of phases, the most important in the present context being those which are magnetic. The interpretation of experimental studies of inversion is made difficult by the fact that inversion may not always go to completion. What follows is a statement of a simplified and possible inversion sequence. The important magnetic phase is the component of the inversion product having spinel structure. This will be a stable near-stoichiometric titanomagnetite. For titanomaghemite with low oxidation state the composition and quantity of this phase is limited by the amount of Fe^{3+} available to form it. Inversion produces a titanomagnetite/ilmenite intergrowth similar to the magnetite/haematite intergrowth produced by inversion of a magnetite/maghemite solid solution. No change in bulk composition takes place during the structural reorganization of inversion so that the coordinates (x, z) of the intergrowth are the same as the pre-inversion titanomaghemite. The titanomagnetite therefore lies at the end of a tie-line passing through the ilmenite compositional point and the point (x, z) and intersecting the titanomagnetite line (Fig. 2.7). The composition (and quantity) of the titanomagnetite so produced can readily be determined. The titanomagnetite inversion product will have composition parameter $x_i = \{x - (3z/2)(1 + x)\}/\{1 - z(1 + x)\}$. The limit for production of this configuration is reached when the titanomaghemite lies on the magnetite-ilmenite line, $z = (2x/3)/(1 + x)$ when x_i becomes zero. Beyond this point there is adequate Fe^{3+} to form magnetite but a deficiency of Fe^{2+} to combine with the TiO_2 to form

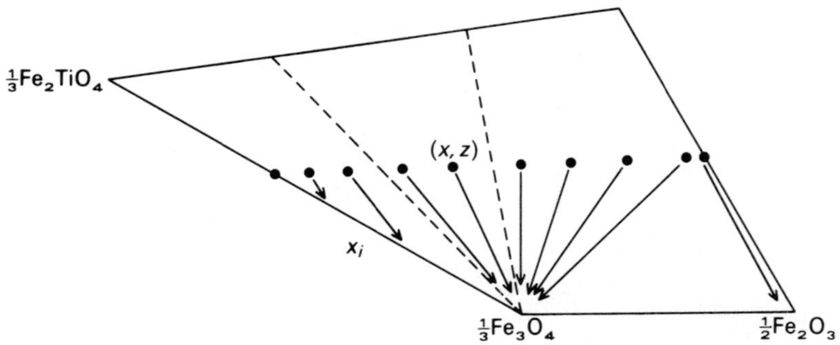

Figure 2.7 A simplified model for the structural inversion of the metastable titanomaghemites to a stable configuration. The magnetic component of the stable intergrowth of phases is indicated. This is a titanomagnetite for a pre-inversion titanomaghemite with composition co-ordinate z up to a certain limit. Above this limit the magnetic phase is magnetite accompanied by other non-magnetic phases, the quantity of magnetite reducing to zero as z approaches unity.

ilmenite. Free TiO_2 will therefore be present in the intergrowth. The limit of this second regime comes at the line joining Fe_3O_4 and TiO_2, $z = (4x/3)/(1+x)$. At this stage there is an excess of Fe^{3+} over Fe^{2+} to form magnetite and the Fe^{2+} limits the quantity of magnetite formed. Fe^{3+} is now found in non-spinel phases (Fe_2O_3, possibly Fe_2TiO_5).

2.1.3 *Haematite and the haematite-ilmenite solid solution*

Haematite, $\alpha - Fe_2O_3$, crystallizes in the corundum structure in which close packed layers are stacked to form a hexagonal close packed lattice (Fig. 2.8). The oxygen framework encloses interstitial sites of six-fold symmetry, the sites forming close packed layers intermediate between the oxygen (0001) layers. In the corundum structure 2/3 of the interstices are filled by cations, the cation layers forming open hexagonal meshes. The cation layers are stacked so that the mesh of the second layer is complementary to the first, i.e. half of the nodes of the second layer lie above the nodes (occupied sites) of the first and half lie above the open centres (vacant sites) of the mesh of layer 1 (Fig. 2.9). Layer 3 similarly complements layer 2, half the nodes of layer 3 lying above the nodes of layer 2 which do not overlie the nodes of layer 1 and half over the open centres of layer 2. Layer 1 is then repeated. Each cation therefore has, in the corresponding sites in the adjacent cation layers on either side, on one side a vacant site and on the other a cation. The sequence along a trigonal axis (containing cations) normal to the layers is: cation : OL : cation : OL : □ : OL : cation : OL : □ and so on (OL = oxygen layer). The hexagonal unit cell of the structure (three times the volume of the rhombohedral unit cell) contains 12 cations and 18 anions. All cation sites are equivalent and have distorted octahedral symmetry. This arises due to electrostatic repulsion between cations in the

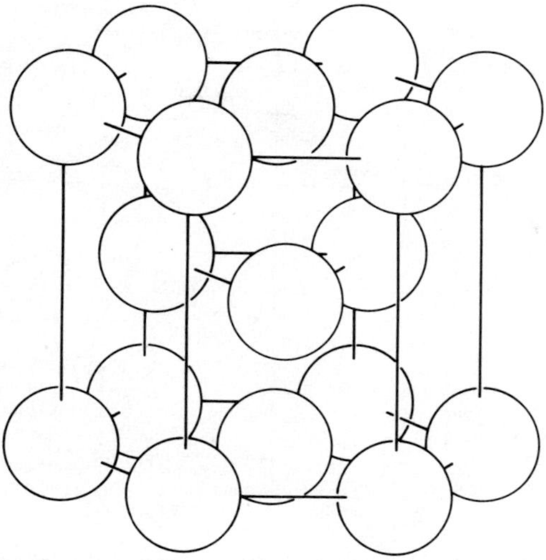

Figure 2.8 The hexagonal close packed structure. The cell contains the equivalent of 6 atoms or ions and an equal number of interstices with six-fold coordination.

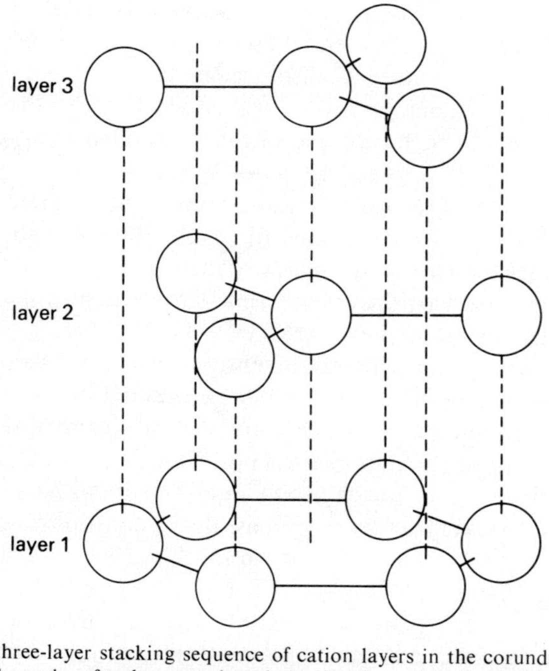

Figure 2.9 The three-layer stacking sequence of cation layers in the corundum structure, the cations lying at the nodes of an hexagonal mesh. Half of the nodes of layer 2 lie above the nodes of layer 1 and half above the vacant sites of layer 1. Layer 3 has the same relation to layer 2, the nodes of layer 2 overlain by the nodes of layer 3 being those which do not overlie nodes of layer 1.

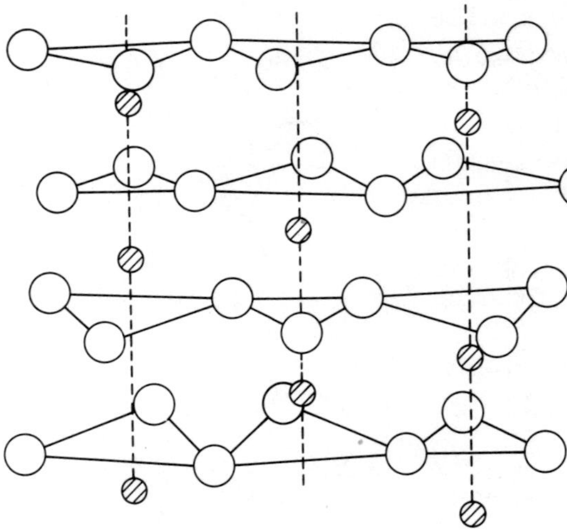

Figure 2.10 The corundum structure showing the sequence of cations (small hatched circles) along a trigonal axis (broken line). The contraction and expansion of oxygen triangles and the displacement of cations from the median planes are indicated.

corresponding sites of adjacent layers. The anion triangles which enclose the axes between the nearest neighbour cations, attracted by the positive potential, contract (Fig. 2.10) and adjacent triangles expand. Half the cations lie above and half an equal distance below the median plane between the oxygen layers. Special position parameters describe the positions of cations and anions. The lattice parameters of the haematite hexagonal unit cell are $a_0 = 5.0345$ Å and $c_0 = 13.749$ Å.

The haematite-ilmenite solid solution may be written $Fe_{2-y}Ti_yO_3$ $(0 < y < 1)$ with $(2 - 2y)$ Fe^{3+} and y Fe^{2+} pfu. The ilmenite structure is essentially similar to the corundum structure but with alternate cation layers occupied by the two different species of cation. In ilmenite the sequence along a trigonal axis will be Fe^{2+} :OL: Ti^{4+} :OL:□ :OL: Ti^{4+} :OL: Fe^{2+} : OL:□.... The hexagonal unit cell dimensions are $a_0 = 5.0881$ Å and $c_0 =$ 14.080 Å. Intermediate members may have ordered or disordered cation arrangements depending on thermal history. In the disordered arrangement the cation layers have the population $(1 - y)$ Fe^{3+}, $y/2$ Fe^{2+} and $y/2$ Ti^{4+} pfu, the ordered arrangement corresponding to layers with $(1 - y)$ Fe^{3+} and y Ti^{4+} adjacent to layers containing $(1 - y)$ Fe^{3+} and y Fe^{2+} pfu.

2.1.4 *The pyrrhotites*

The pyrrhotites may be represented by the formula $Fe_{1-x}S$ where $0 < x < 0.13$ although only values of x corresponding to certain rational Fe:S ratios appear to be possible. The valency of the cation species for non-zero x seems indeterminate. It may be that some of the cations are Fe^{3+} or at least the average charge per Fe ion is greater than + 2. Alternatively, the S ions may have an average negative charge of less than 2 with adjacent S ions sharing electrons to form stable molecular ions. The pyrrhotites have structures based on the NiAs structure. This may be derived from the hexagonal unit cell of Fig. 2.8 by adding a second hexagonal prism above or below, the three sites within the body of the second prism being in complementary positions to those in the first. In stoichiometric troilite, FeS, the sites within the bodies of the prisms are occupied by six S^{2-} and sites on the surfaces by Fe^{2+}, both ions in six-fold coordination with nearest neighbours. A smaller, primitive, unit cell, containing only two FeS formula units, is provided by a trigonal prism one-third of the volume of the hexagonal NiAs prism.

The non-stoichiometric pyrrhotites Fe_7S_8 (monoclinic pyrrhotite) Fe_9S_{10} and possibly $Fe_{10}S_{11}$ and $Fe_{11}S_{12}$ (low temperature hexagonal pyrrhotite) are cation deficient. In Fe_7S_8 it is believed that one quarter of the sites in alternate cation (0001) layers are occupied by vacancies. This leads to an orthorhombic unit cell containing eight Fe_7S_8 formula units with an alternative and smaller monoclinic unit cell. In the more stoichiometric compounds it may be that certain cation layers contain one quarter vacancies and are separated by sequences of one or two filled layers. The phase relations and structural properties of the pyrrhotites are complicated with possible transformations between ordered and disordered cation/vacancy arrangements, and from single phase to multiphase intergrowths occurring above room temperature.

2.1.5 *Other iron-bearing minerals*

Numerous minerals in all categories contain iron and are of interest either as sources of magnetic minerals or as the alteration products of such minerals. Here we shall consider the mineral species of each category which are richest in iron and are of some importance in the magnetopetrological history of rocks.

(i) *Non-silicates.* The commonest form of iron oxyhydroxide and the most stable is *goethite*, α-FeOOH. The dimensions of the orthorhombic unit cell are: $a = 4.59$, $b = 9.95$ and $c = 3.02$ Å. As in the case of haematite, the structure contains an hexagonal close packed oxygen lattice and on heating to about 200–400°C goethite dehydrates to haematite. *Lepidocrocite*, γ-FeOOH, dehydrates to form maghemite at temperatures between 200 and 250°C. The orthorhombic unit cell of lepidocrocite has dimensions $a = 3.87$, $b = 12.51$ and $c = 3.06$ Å. *Akaganeite*, β-FeOOH, with tetragonal unit cell of dimensions $a = 10.48$ and $c = 3.02$ Å dehydrates at about 250°C to form haematite possibly via an intermediate phase which facilitates the drastic reorganization of the crystal structure needed for the akaganeite-haematite transformation. *Siderite*, $FeCO_3$, or solid solutions with magnesite, $MgCO_3$, or other carbonates, decomposes to form an iron oxide or oxyhydroxide depending on the conditions during alteration. Pure siderite crystallizes in the trigonal system with a rhombohedral unit cell $a = 5.8$, $\alpha = 47°$. *Pyrite*, FeS_2, is cubic with cell edge 5.42 Å.

(ii) *Silicates.* Of *orthosilicates*, the ferro-magnesian olivines, the solid solution between fayalite, Fe_2SiO_4, and forsterite, Mg_2SiO_4, crystallize in the orthorhombic system with unit cell dimensions ranging from $a = 4.82$, $b = 10.48$, $c = 6.10$ (fayalite) to $a = 4.76$, $b = 10.20$, $c = 5.98$ (forsterite). Compositions having approximately 20% fayalite are common in basic igneous rocks. High temperature oxidation of the olivines produces magnetite. During low temperature hydrothermal alteration, metamorphism or weathering, magnetite, haematite, goethite and other iron-bearing alteration products, chlorite and amphiboles are produced. Among *chain silicates*, orthopyroxenes, the solid solution between ferrosilite, $FeSiO_3$, and enstatite, $MgSiO_3$, may be altered forming iron oxides, oxyhydroxides and other iron-containing minerals such as the iron-rich amphiboles typified by the grunerite/cummingtonite series $(Fe, Mg)_7(Si_8O_{22})(OH)_2$ which may in turn be altered to produce iron oxides. The following *sheet silicate* minerals are potential or actual sources of magnetic phases:

biotite, $K_2(Mg, Fe^{2+})_{6-4}(Fe^{3+}, Al, Ti)_{0-2}(Si, Al)_8O_{20}(OH, F)_4$
chlorite $(Mg, Al, Fe)_{12}(Si, Al)_8O_{20}(OH)_{16}$
clay minerals, e.g. smectites

$$(\tfrac{1}{2}Ca, Na)_{0.7}(Al, Mg, Fe)_4(Si, Al)_8O_{20}(OH)_4nH_2O.$$

2.2 The magnetic mineralogy of igneous rocks

The composition and constituents of the magnetic mineral assemblage developed in an igneous rock depend on the bulk composition of the magma from which the minerals crystallize, the rate of cooling, the composition of magmatic gases and access to the atmosphere (or ocean). The microstructure of the magnetic minerals—the porosity, size, shape and orientation of the grains together with their geometrical relation to other phases—is also determined by these factors. These factors vary locally, sometimes systematically with position, in an igneous body. The number of degrees of freedom in such a system is large and prohibits systematic investigation. It is, nevertheless, possible to give a simplified general account of the various paths along which the magnetic mineral fraction in an igneous rock may travel from the point of crystallization to its state when collected for study.

The bulk composition of the magma dictates the order in which the minerals crystallize. The oxygen coordination polyhedra surrounding cations are present in the melt, the metal-oxygen distances being greater than in the corresponding crystalline state. The electrostatic field due to the neighbouring anions therefore produces a smaller splitting of the electronic 3*d* orbitals in the melt. The greater splitting in the solid state provides deeper energy levels and a lower energy for certain 3*d* orbital electron populations (see section 3.1.). This may explain why Cr^{3+} has a strong affinity for the solid phase, chromian spinels crystallizing early, before the Fe–Ti oxides, occasionally producing a titanomagnetite with a chromium-rich core. The compositions of the first-formed near-stoichiometric titanomagnetite and members of the haematite-ilmenite solid solution also depend on the bulk chemistry of the magma. In the case of a 'basic' basalt the titanomagnetite composition lies in the range $0.5 < x < 0.8$ and the haematite-ilmenite in the range $0.8 < y < 1.0$. Together they constitute 1–10% by volume of the rock. The titanium content of the oxides is lower in 'acid' and 'intermediate' rocks. The various possible paths by which the oxide system evolves during cooling may be described using the phase diagram of the titanomagnetite solid solution (Fig. 2.11). The diagram is based on studies of simple synthetic materials and is an approximation of the characteristics of the more complicated natural multicomponent titanomagnetites. The point A on the liquidus represents the composition of the liquid in equilibrium with the first-formed solid for a bulk composition in the binary system defined by $x = 0.6$. The composition of the liquid is richer in titanium than the solid, the composition of which lies on the intersection of the isotherm through A

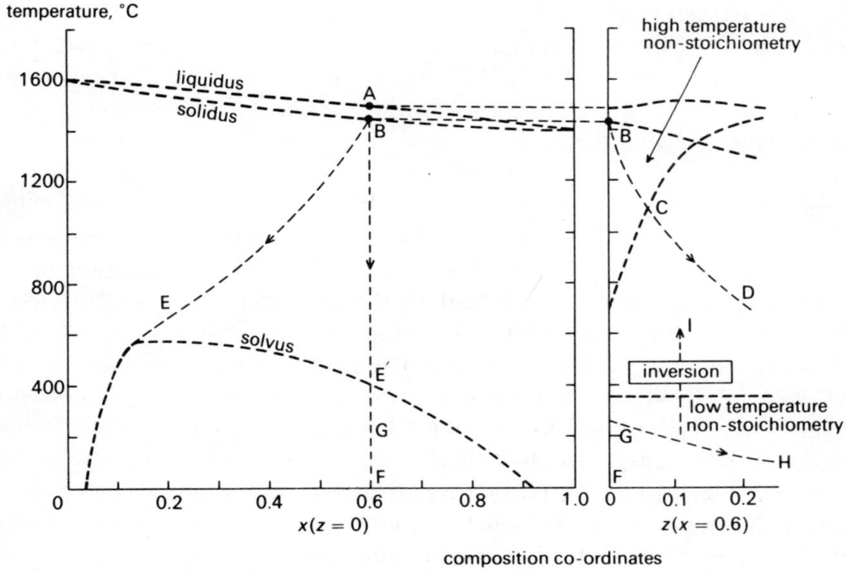

Figure 2.11 The phase diagram for the titanomagnetite binary solid solution and for a section along an oxidation line $x = 0.6, 0 < z < 1$.

and the solidus. As temperature falls further the compositions of the co-existing liquid and crystallizing solid lie at the intersections of horizontal tie-lines with the liquidus and solidus curves. The bulk composition remains on the isopleth although points on the section *AB* do not represent actual states of the system, the zone between the liquidus and solidus being a void in which no homogeneous phase exists. The liquid becomes richer in titanium as does the bulk composition of the crystallizing solid as the temperature falls. The point *B* represents the final bulk composition of the solid and the corresponding point on the liquidus the composition of the last drop of liquid. The crystals of the solid phase are therefore zoned having iron-rich cores. Diffusion rates are high enough that the zoning will be removed unless rapid quenching takes place through the liquidus/solidus region. It may be that heterogeneities arise even in slowly crystallized titanomagnetite due to growth around pre-existing crystals of chromite or primary ilmenite. Iron-rich cores in titanomagnetite/titanomaghemite grains of submarine basalts are probably not a relic of the rapid cooling but due to low temperature alteration (see below).

The path followed by the solid, as temperature falls further, depends on the rate of cooling and the oxygen pressure in the system. In general the system will be oxidizing with respect to the stoichiometric titanomagnetite, the free oxygen being produced by the dissociation of magmatic water vapour, the hydrogen having escaped due to its ability to diffuse rapidly through the rock matrix. The extent to which the Fe–Ti oxide is able to approach equilibrium (becomes 'deuterically' oxidized) depends on the time available and the local oxygen pressure. A possible path by which the degree of oxidation may develop with time, and falling temperature, is shown by the curve *BCD* in the right hand side of Fig. 2.11. The first part of the curve corresponds to a non-stoichiometric titanomagnetite, the composition lying in the plane of the figure. Beyond the point *C* a single phase compound cannot be maintained and the curve *CD* represents the bulk composition of an intergrowth of phases, a titanomagnetite richer in iron than $x = 0.6$ coexisting with ilmenite. (This is essentially similar to the products of the inversion of titanomaghemites described above, the inversion following on the ascent of an isopleth towards the curve *CD*, leaving the zone of low temperature non-stoichiometry.) The composition of the spinel phase is indicated by the curve *BE* in the left-hand side of Fig. 2.11. The original titanomagnetite crystal becomes subdivided by lamellae of ilmenite lying in the (111) planes of the spinel host. The quantity and composition of the coexisting phases, as oxidation proceeds along the $x = 0.6$, $0 < z < 1$, path may be similar to the sequence discussed in the context of the inversion of titanomaghemites of the same composition. The sequence for the spinel phase is therefore first of all a progressive iron-enrichment until the magnetite composition is reached (as in Fig. 2.7). Beyond this point the spinel phase is magnetite, coexisting with non-magnetic phases, probably ilmenite and rutile (TiO_2).

An excess of Fe^{3+} over that required to form magnetite results from further oxidation and Fe^{3+} occurs in non-spinel phases, possibly pseudobrookite, Fe_2TiO_5, if the temperature is high enough. As the Fe^{2+} concentration falls to zero, so does the possible concentration of magnetite. Ultimately, haematite is the stable magnetic phase.

The deuteric oxidation of the magnetite or titanium-poor titanomagnetite of acid rocks is essentially a similar process to that for the $x = 0.6$ titanomagnetite, but simplified by the absence of titanium. Lamellae of haematite subdivide the magnetite, the magnetite eventually being completely replaced by haematite ('martitization').

Returning now to the freshly crystallized titanomagnetite of point *B* we may consider the case of rapid cooling. This will result in a single-phase near-stoichiometric titanomagnetite (point *F*), there being inadequate time for oxidation to take place. The solvus curve for the binary titanomagnetite intersects the $x = 0.6$ isopleth at the point E'. Prolonged slow cooling (and a non-oxidizing environment) between E' and *F* would result in exsolution of the titanomagnetite to compositions lying at the intersections of isotherms and the solvus. Eventually an intergrowth of two compositions, $x \sim 0.05$ and 0.95, is produced, the near-ulvöspinel phase exsolved parallel to the (001) planes of the magnetite.

Slow cooling below about 400°C, or mild reheating in an oxidizing environment, may produce oxidation (along G/F to *H*) within the low temperature zone of non-stoichiometry (maghemitization). As in the case of deuteric oxidation, the degree of development of the oxidized state depends on the interplay of reaction and cooling rates, and equilibrium states are not necessarily reached. Diffusion rates are slower at the lower temperatures and homogenization within the grains may proceed more slowly than the surface oxidation. Larger grains will have more pronounced concentration gradients, smaller grains may be essentially homogeneous, depending on the time-temperature cooling regime. In the submarine environment maghemitization may take place at ambient temperature by the leaching of iron by seawater. In this case the spinel lattice contracts to accommodate the reduced volume of cations. As the grains are intergrown with and bound to the matrix, especially for the skeletal forms of the rapidly crystallized titanomagnetites of submarine basalts, strains are set up which result in a splitting of the grains. Splitting or cracking within the body of the grain suggests maghemitization throughout the grain although not necessarily to a uniform degree. Splitting and cracking in skins imitating the form of the surface of the grain would suggest greater volume contraction and degree of alteration at the surface. The iron leached out by seawater, or during supergene weathering generally, will rapidly oxidize to produce rust or some kind of crystalline or amorphous oxyhydroxide. Elevated temperatures (above about 350°C) bring about the inversion of the titanomaghemite (*H* to *I* in Fig. 2.11).

Other iron-bearing minerals may also be altered from their post-crystallization state during cooling. Primary ilmenite will deuterically oxidize just as the ilmenite component of a deuterically oxidized titanomagnetite oxidizes during the later stages of oxidation. Haematite or a haematite-rich member of the haematite-ilmenite series results, together with TiO_2 and possibly Fe_2TiO_5. Iron-bearing silicates, the olivines, and pyroxenes are susceptible to deuteric oxidation during the cooling of an igneous body. The olivines can accommodate only a limited concentration of Fe^{3+} in a defect olivine structure before breaking down to form magnetite inclusions among more silica-rich phases. The magnetite may be further oxidized to haematite. Pyroxenes and other silicates, plagioclase, biotite and amphibole may contain iron oxide inclusions.

Iron oxides and oxyhydroxides are also produced during low-temperature alteration of silicates during the processes of serpentinization and iddingsitization.

2.3 The magnetic mineralogy of sediments

The physical and chemical weathering and erosion of primary igneous rocks provides the material for sedimentary rocks. The sedimentary rocks may, in turn, be eroded to provide the material for further sediments. Sedimentary rocks may contain any of the magnetic mineral phases of igneous rocks. Alteration of the magnetic minerals during weathering, transport and sedimentation may produce a mineral assemblage in a highly oxidized state. The ultimately stable iron oxide, haematite, can be the principal magnetic constituent of sediments. Non-magnetic minerals in the source rocks may also be the origin of haematite (or magnetite) in sediments, either by direct transformation or via iron-bearing solutions.

The characteristic times for the physicochemical evolution of the magnetic mineral assemblage in sediments are long compared to the laboratory time scale. Simulation of the chemical processes believed to occur in sediments cannot necessarily be accelerated by, for example, operating at elevated temperatures. The available picture for the genesis and evolution of the magnetic mineral fraction in sediments is therefore less well established than is the case for igneous rocks.

Red sandstones have proved to be a particularly useful carrier of the palaeomagnetic record. The term covers a variety of sediments with colours ranging from brown to purple. A common characteristic is the pigmentation of finely divided haematite or ferric oxyhydroxide. Red sandstones or red beds may be derived from the erosion of the red soils of moist tropical areas and hot deserts, these soils containing amorphous ferric hydroxide (limonite), ferric oxyhydroxide (principally goethite) and ferric oxide. Subsequent dehydration, after deposition, may enrich the sediment in haematite. The progressive breakdown of iron-bearing silicates may also produce pigment. In addition to the finely-divided red pigment, haematite is also present in red beds in the form of black grains of specularite, possibly resulting from the martitization or maghemitization/inversion of detrital magnetite particles. Unaltered magnetite is sometimes present but forms a progressively lower concentration of the magnetic fraction with increasing age of the sediment.

The magnetic mineralogy of many of the drab sediments which are finding application in palaeomagnetism, in some cases due to the improved performance of remanence-measuring magnetometers, is even more elusive. In sediments which have an organic element in their genesis, such as limestones, deep-sea sediments and lake sediments, a source of remanence carriers may be small magnetite particles contained in the nervous systems of primitive animals. The alteration of the magnetite particles within the living animal may be inhibited by an inert coating around the grains which may also prevent subsequent alteration in the sediment.

2.4 The magnetic mineralogy of extra-terrestrial materials

Whereas the magnetic phases of the rocks of the Earth's crust are compounds of iron, in the case of extra-terrestrial material—samples of lunar rock and dust and meteorites—the magnetic phases are metallic iron or iron-nickel alloys. The conditions under which lunar rocks crystallize are evidently much more reducing than for terrestrial rocks. Thus Fe^{3+} concentrations in the oxides are extremely low, and the iron-rich end members of lunar spinel oxides are the Fe^{2+}-bearing chromite ($FeCr_2O_4$), hercynite ($FeAl_2O_4$) and ulvöspinel. In some oxide minerals, titanium occurs in the trivalent form. The magnetic phase is body-centred cubic α-iron, occurring in a wide range of grain sizes. The magnetic mineralogy of lunar igneous rocks is therefore much simpler than their terrestrial equivalents. Variations in conditions of crystallization and cooling rates, although producing variations in composition and zoning in the non-magnetic spinel oxides, affect only the concentration and microstructure (essentially only grain size) and not the composition of the magnetic phase.

Meteorites contain iron-nickel alloys in concentrations from 100% (iron meteorites) to about 1% (achondrites). Body-centred cubic (bcc) α-iron can accommodate only a limited concentration of Ni (about 5 atomic %). Higher concentrations require the fcc lattice. This is also the structure of nickel itself and the high temperature form of iron (the γ form, at temperatures above 900°C). The bcc alloys are called kamacite and the fcc, taenite. Subsolvus exsolution of taenite produces an intergrowth of an iron-rich kamacite and a nickel-enriched taenite.

2.5 Summary

The magnetic minerals of the Earth's crust are compounds of iron, commonly oxides. The magnetic mineral of primary igneous rocks is titanomagnetite which crystallizes in the spinel structure having approximate composition $Fe_{3-x}Ti_xO_4 (0 < x < 1)$. In basic rocks $x \sim 0.6$ and has lower values in acid rocks.

Deuteric oxidation may take place within the cooling igneous body transforming the primary titanomagnetite to an intergrowth of phases which includes magnetite $Fe_3O_4 (x \sim 0)$. Otherwise more rapid cooling preserves the titanomagnetite intact.

Low temperature oxidative alteration ('maghemitization') of the titanomagnetites produces titanomaghemite, a non-stoichiometric oxide with spinel structure. The evolving path of the mineral depends to what extent oxidation takes place by addition of oxygen or removal of iron. The composition of titanomaghemites may be conveniently expressed $Fe_{(3-x)R}Ti_{xR}$-$\square_{3(1-R)}O_4$ where the parameter R falls from 1 to a minimum of $8/(9+x)$ during maghemitization. An oxidation parameter z (the 'fraction of reaction') may also usefully be introduced. The titanomaghemites are structurally metastable and 'invert' without change in bulk chemical composition, a simple example of inversion being the transformation of maghemite, γ–Fe_2O_3, to haematite, α–Fe_2O_3.

Haematite crystallizes in the corundum structure which, like the cubic spinel structure, is constructed of stacked close packed oxygen layers, the stacking sequence producing an hexagonal close packed lattice. Being the ultimately stable, most highly oxidized, magnetic phase, haematite is the principal remanence carrier in sediments, notably the red sandstones, in which it occurs as pigment and as grains of specularite.

Other primary iron-bearing minerals, such as ilmenite $FeTiO_3$ and olivines, are non-magnetic but may be altered during cooling, or at a later time to produce magnetic phases, principally magnetite and haematite. The primary magnetic minerals may also be ultimately altered to non-magnetic phases. The iron-bearing minerals of the Earth's crust may undergo many changes in composition and form in the course of geological history.

Selected bibliography

Akimoto, S. (1962) Magnetic properties of FeO–Fe_2O_3–TiO_2 system as a basis of rock magnetism. *J. Phys. Soc. Japan* **17** (Suppl. B1), 706–710.

*Haggerty, S.E. (1976) 'Oxidation of opaque mineral oxides in basalts', in *Oxide Minerals*, (ed D. Rumble III) Mineralogical Society of America, Washington, D.C., Chapter 4.

Lindsley, D.H. (1961) Investigation in the system FeO–Fe_2O_3–TiO_2. *Carnegie Inst. Washington Yearbook* **61**, 100–106.

*Lindsley, D.H. (1976) 'The crystal chemistry and structure of oxide minerals as exemplified by the Fe–Ti oxides', in *Oxide Minerals*, (ed. D. Rumble III) Mineralogical Society of America, Washington, D.C., Chapter 1.

O'Reilly, W. (1965) A low temperature study of synthetic titanomagnetites and controlled oxidation products. Ph.D. Thesis, University of Newcastle upon Tyne.

*Ryall, P.J.C. and Hall, J.M. (1980) Iron loss in titanomagnetite during low temperature oxidation. *J. Geomag. Geoelectr.* **32**, 661–669.

O'Reilly, W. and Readman, P.W. (1971) The preparation and unmixing of cation-deficient titanomagnetites. *Z. Geophys.* **37**, 321–327.

*Petersen, N., Eisenach, P. and Bleil, U. (1979) 'Low temperature alteration of magnetic minerals in ocean floor basalts', in *Deep Drilling Results in the Atlantic Ocean: Ocean Crust*, (ed. Talwani *et al.*) Lithocrafters Inc., Chelsea, Michigan, 169–209.

Readman, P.W. (1972) A magnetic study of non-stochiometric titanomagnetites. Ph.D. Thesis, University of Newcastle upon Tyne.

Syono, Y., (1965) Magnetocrystalline anisotropy and magnetostriction of Fe_3O_4–Fe_2TiO_4 series—with special application to rock magnetism. *Japan. J. Geophys.* **4**, 71–143.

*van Houten, F.B. (1968) Iron oxides in red beds. *Geol. Soc. Amer. Bull.* **79**, 399–416.

*Ward, J.C. (1970) The structure and properties of some iron sulphides. *Rev. Pure Appl. Chem.* **20**, 175–206.

*Sources not specifically cited in text.

3 The atomic basis of magnetism

3.1 The electron spin; the transition elements

In the large-scale world we commonly associate magnetic fields with electrical currents. Our experience in the large-scale world is of little value, however, in providing a theoretical description of the origin of the magnetism of remanence-carrying materials. Although we can use our experience of classical physics to equate an orbiting electron to a current loop and determine the resultant magnetic dipole moment, this turns out to be of little help, as the observed magnetism of solids is due almost exclusively to the magnetic dipole moment associated with the 'spin' of the electron. The spin of the electron has no analogue in classical physics and is a feature of the quantum mechanical description of the inhabitants of the submicroscopic world of atoms and electrons. The electron has angular momentum quantized in units of $h/2\pi$ $(=\hbar)$ where h, Planck's constant, equals 6.63×10^{-34} joule s, and a spin quantum number $s = \frac{1}{2}$. The component of spin angular momentum in a specified direction is $s\hbar$ and the associated magnetic moment $(e/m)s\hbar$ where e and m are the electronic charge and mass. The electron can therefore be regarded simply as a microscopic magnet with magnetic moment $(e\hbar)/(2m)$, this being the fundamental unit of magnetic moment, the Bohr magneton, $\beta(= 9.27 \times 10^{-24}$ A m^2).

Quantum mechanics provides us with a picture of the electronic structure of the atom in which the electrons reside in states identified by labels n, l, m_l (which take integral values) and s. l may take values up to $(n - 1)$, and m_l has values 0, $\pm 1, \ldots, \pm l$. Thus for $n = 1$, the lowest energy state, $l = 0$ (the '1s state'); for $n = 2$, $l = 0$ (the 2s state) and $l = 1$ (the three 2p states for which $m_l = 0$ and ± 1); for $n = 3$, $l = 0$ (the 3s state), $l = 1$ (the three 3p states, $m_l = 0$ and ± 1) and $l = 2$ (the five 3d states, $m_l = 0$, ± 1 and ± 2). For the present purpose the quantum numbers can simply be regarded as identifying labels which are subject to certain numerical rules. They do, of course, have physical significance—n describing the energy of the electron in the state, l the total orbital angular momentum, m_l the size of a particular component of orbital angular momentum and s the spin angular momentum. The reader is referred to any undergraduate atomic physics text (e.g. Littlefield and Thorley, 1979) for an account of quantum numbers and the electronic structure of atoms.

A physical picture of the states is obtained by plotting surfaces on which there is a constant probability of finding the electron (the electronic orbitals).

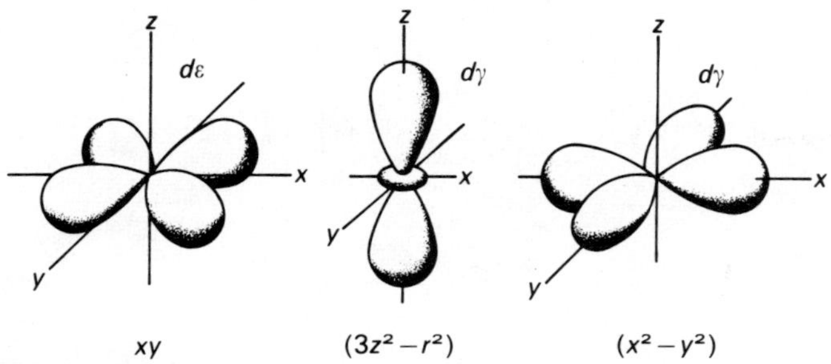

Figure 3.1 The representation of the 3*d* electronic states as orbitals. There are five 3*d* orbitals, two of which are not shown in the figure. These are similar to the left-hand ($d\varepsilon$) orbital in which the lobes are shown as lying along diagonals in the xy plane. The two $d\varepsilon$ orbitals not illustrated lie in the yz and zx planes.

The 3*d* orbitals are shown in Fig. 3.1. In constructing the table of the elements, states are progressively filled with electrons following two rules—the Pauli principle and the Hund rule of maximum spin multiplicity. The Pauli principle decrees that no two electrons in the same atom may have the same set of values of the four quantum numbers. Two electrons in the same orbital (the same values of n, l and m_l) therefore have different s, one with spin 'up' and one with spin 'down' (and zero net magnetic moment). Within a given subshell (e.g. the 3*d* states) the electron spins are combined to give maximum resultant spin vector according to the Hund rule, subject to the Pauli principle not being violated. Table 3.1 shows the operation of this process for that part of the periodic table of the elements between Na and Zn. This includes the first transition series of elements and other species common in minerals. As each subshell is progressively filled the resultant spin vector and spin magnetic moment rises to a maximum value for a half full subshell and falls to zero for a fully occupied subshell. The resultant spin quantum number, S, of the 3*d* subshell rises to a maximum of $\frac{5}{2}$, no contribution coming from the 18 electrons of the argon core, $1s^2 2s^2 2p^6 3s^2 3p^6$. It can be seen from the table that the 4*s* levels are filled before the 3*d* levels are occupied and later partially emptied to provide a half full ($3d^5$) and completely filled ($3d^{10}$) lower energy subshell.

In the case of oxide mineral compounds, it is the ionic species of the elements which are of interest. In ionizing the transition metals, the 4*s* electrons are removed first. Iron occurs commonly as Fe^{2+} ($3d^6$) and Fe^{3+} ($3d^5$) ions which may be regarded simply as microscopic magnets ('paramagnetic ions') with spin magnetic moments of 4β and 5β respectively (in the 'high-spin' state, see below). Ti^{4+} ($3d^0$) has no magnetic moment (a 'diamagnetic ion'). Oxygen has an electron structure $1s^2 2s^2 2p^4$ and on acquiring two electrons to become O^{2-} achieves the filled shell Ne structure with zero magnetic moment. The same filled shell is found for Si^{4+}, Al^{3+} and Mg^{2+}. Mn^{2+} ($3d^5$, 5β) and Cr^{3+} ($3d^3$, 3β) are paramagnetic species occurring in small quantities in magnetic minerals.

Table 3.1 The electron structure of the elements, Na to Zn

		3s	3p			3d					4s
Na		↑									
Mg		↑ ↓									
Al			↑								
Si			↑	↑							
P			↑	↑	↑						
S			↑ ↓	↑	↑						
Cl			↑ ↓	↑ ↓	↑						
A			↑ ↓	↑ ↓	↑ ↓						
K											↑
Ca	$1s^2 2s^2 2p^6$										↑ ↓
Sc		$3s^2$				↑					↑ ↓
Ti						↑	↑				↑ ↓
V						↑	↑	↑			↑ ↓
Cr			$3p^6$			↑	↑	↑	↑	↑	↑
Mn						↑	↑	↑	↑	↑	↑ ↓
Fe						↑ ↓	↑	↑	↑	↑	↑ ↓
Co						↑ ↓	↑ ↓	↑	↑	↑	↑ ↓
Ni						↑ ↓	↑ ↓	↑ ↓	↑	↑	↑ ↓
Cu						↑ ↓	↑ ↓	↑ ↓	↑ ↓	↑ ↓	↑
Zn						↑ ↓	↑ ↓	↑ ↓	↑ ↓	↑ ↓	↑ ↓

In a free atom or ion, the 3*d* states are of equal energies (they are degenerate). This is no longer so for the ion incorporated in an oxide crystal lattice, which experiences the electrical field (the crystal field) due to the nearest neighbour ions. The symmetry of the field depends on the symmetry of the array of charged nearest neighbours. In octahedral symmetry the six nearest neighbours are most simply regarded as lying on the principal axes of a cartesian coordinate system. The four nearest neighbours in tetrahedral symmetry are similarly assigned to the principal diagonals of a cartesian system. The energies of electrons in the 3*d* states now depend on the way the lobes of the orbitals fit into the symmetry, the result being a division of the states into two groups, one termed $d\varepsilon$ (or t_{2g}) and the other $d\gamma$ (or e_g). This is shown in Figs. 3.2 and 3.3. Electrons in the lobes of the

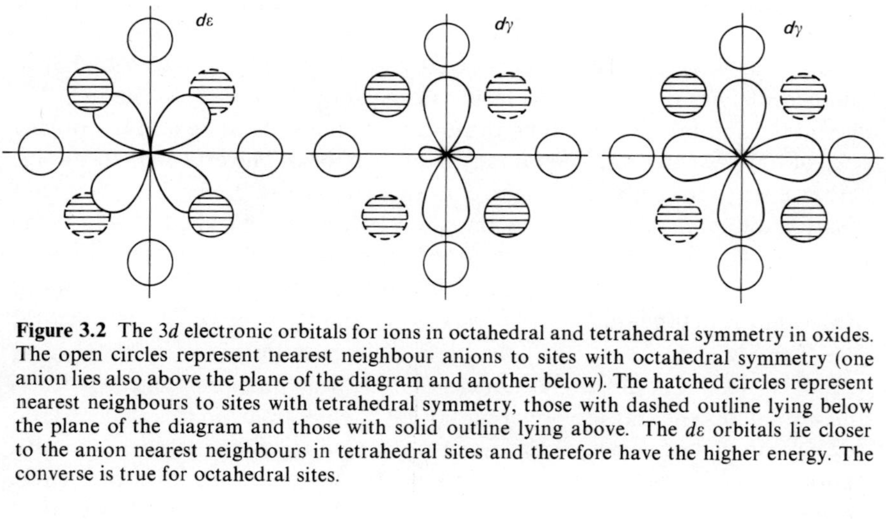

Figure 3.2 The 3*d* electronic orbitals for ions in octahedral and tetrahedral symmetry in oxides. The open circles represent nearest neighbour anions to sites with octahedral symmetry (one anion lies also above the plane of the diagram and another below). The hatched circles represent nearest neighbours to sites with tetrahedral symmetry, those with dashed outline lying below the plane of the diagram and those with solid outline lying above. The $d\varepsilon$ orbitals lie closer to the anion nearest neighbours in tetrahedral sites and therefore have the higher energy. The converse is true for octahedral sites.

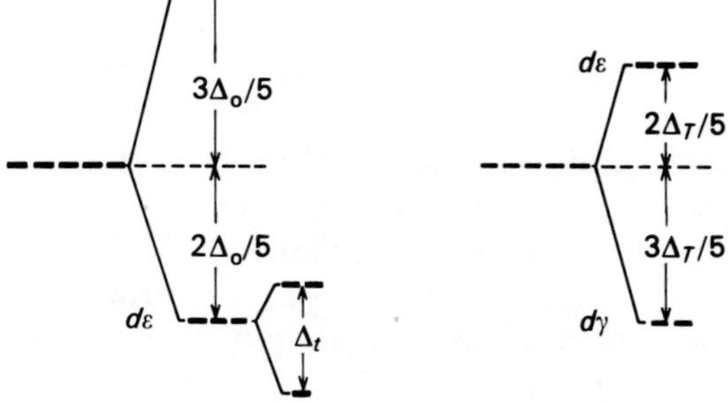

Figure 3.3 The energy level diagrams for electrons in 3*d* states in sites of octahedral and tetrahedral symmetry. The five degenerate states of the free ion are split into an upper doublet ($d\gamma$) lying Δ_0 above a lower triplet ($d\varepsilon$) in the case of octahedral symmetry and an upper triplet ($d\varepsilon$) Δ_T above a lower doublet ($d\gamma$) in tetrahedral symmetry. A further splitting Δ_t in the octahedral case is shown due to the trigonal symmetry of nearest neighbour cations.

three $d\varepsilon$ orbitals, which lie in the xy, yz and zx planes, approach the negatively charged anions in tetrahedral symmetry, whereas those in the two $d\gamma$ orbitals, which lie along the axes, do not. The energies of the $d\varepsilon$ orbitals are therefore raised with respect to the $d\gamma$ orbitals. Inspection of the figure will show that the converse is the case for octahedral symmetry. The five electrons of Fe^{3+} still obey the Hund rule, provided the splitting of the energy levels is not too great, and all $3d$ states are half-filled, electrons occupying the upper energy levels with spin 'up' rather than completely filling lower levels and reducing the resultant spin. The 'high spin' state ($S = \frac{5}{2}$ for Fe^{3+}) obtains in transition ions of the minerals of crustal rocks. The splitting due to the crystal field will be increased if the nearest neighbour anion coordination sphere contracts (due to high pressure) and decreases if the coordination sphere expands (as in the molten state).

The difference in energy of the cation in the free and bound states is called the crystal field stabilization energy. The three $3d$ electrons of Cr^{3+} occupy the three $d\varepsilon$ states (depressed by an amount $\frac{2}{5}\Delta_0$, see Fig. 3.3) in octahedral symmetry. As Δ_0 is larger in the solid state than in the melt the crystal field stabilization will be the greater in the solid state and the preference of Cr^{3+} for the solid phase may be understood. The crystal field stabilization energy may also play a role in describing the preference of certain ions for different symmetries in the solid phase (the tetrahedral and octahedral sites of the spinel structure, for example). If the electron configuration of an ion is $(d\varepsilon)^p\ (d\gamma)^q$, the crystal field stabilization (i.e. energy reduction) of an ion on octahedral sites is $(\frac{2}{5}p - \frac{3}{5}q)\Delta_0$ and $(\frac{3}{5}q - \frac{2}{5}p)\Delta_T$ on tetrahedral sites (see Fig. 3.3). As $\Delta_0 > \Delta_T$ the preference of certain ions for octahedral symmetry (e.g. Cr^{3+}) can be understood. A zero site preference of Fe^{3+} and a small octahedral site preference of Fe^{2+} emerges from this electrostatic description. An extension of this description to one which includes a combination of the oxygen p orbitals and the metal $3d$ orbitals to form molecular orbitals provides a reason for the observed preference of Fe^{3+} for tetrahedral symmetry in spinels.

3.2 Exchange

Although the magnetic moments of the atomic magnets are minute, magnetic materials can be used to powerful and dramatic effect in the large-scale world. This is because when magnetic atoms or ions are incorporated into a crystal lattice they may cooperate to produce an ordered array of magnetic vectors throughout large regions of the lattice. The mechanism for this cooperation is the exchange interaction resulting from the overlap of the orbitals of nearest neighbour atoms or ions. In the case of oxide compounds the oxygen ions provide a link between nearest neighbour cations which are otherwise too far apart for direct overlap of their $3d$ lobes. The mechanism requires quantum mechanics for even a qualitative treatment but can be

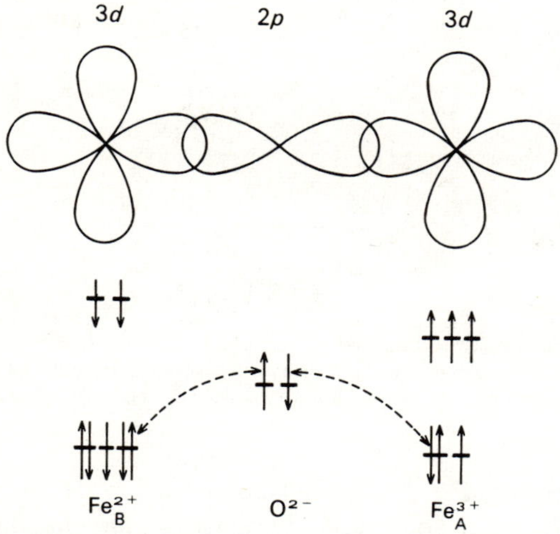

Figure 3.4 The indirect exchange interaction between two cations via an oxygen anion. The mutually antiparallel $2p$ electrons are partly shared by the cations which thus acquire a coherent spin vector orientation, in the illustrated case, antiparallel to each other.

described by a simple model invoking the Pauli principle and Hund's rule of maximum spin multiplicity.

Consider two magnetic atoms separated by an anion as shown schematically in Fig. 3.4. The lobes of one of the oxygen $2p$ orbitals are shown overlapping with the lobes of one of the $3d$ orbitals of the metallic ion. The $2p$ lobe contains two electrons with antiparallel spins, each of which can be imagined as being shared with one of the cations on either side of the combined orbital system. The Pauli principle requires that if the transferred electron occupies a $3d$ state already containing an electron, the spins of the two must be antiparallel. If the transferred electron occupies an empty $3d$ state, its spin will be parallel to those of the electrons in the other $3d$ states of the same ion because of the Hund rule. The resultant spin of one cation will thus become coupled coherently to the spin of the other via the antiparallel spins of the electrons in the $2p$ state. It will be seen that when one cation has a less than half-filled $3d$ subshell and the other has at least a half-filled $3d$ subshell, the resultant spin vectors of the cations are coupled parallel. In all other cases the coupling will be antiparallel. The diagram illustrates the $3d^5$–$3d^6$ (Fe^{3+}–Fe^{2+}) interaction in which the Fe^{3+} and Fe^{2+} are shown in tetrahedral and octahedral sites as they may be in many spinel oxides.

The effectiveness of the interaction depends on the degree of overlap of the orbitals which in turn depends on the metal-oxygen distance and the metal-oxygen-metal angle which should be near 180°. The spin structure which results therefore depends on both the type of cation and the crystal structure. A two-dimensional representation of a simple antiferromagnetic

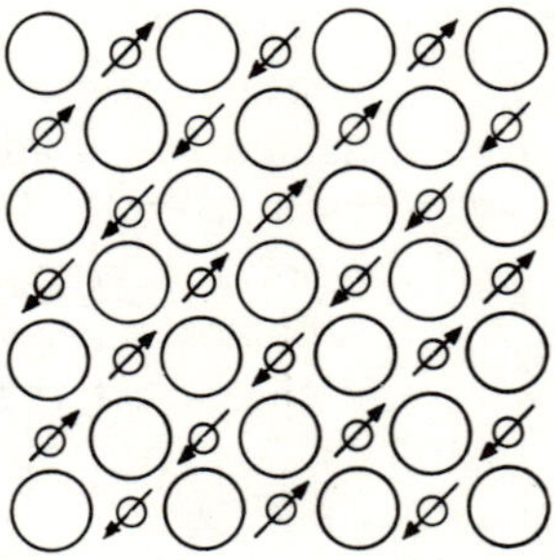

Figure 3.5 A schematic two-dimensional representation of a simple antiferromagnetic monoxide. Nearest cation neighbours linked by oxygens have antiparallel spin vectors and the spin structure is made up of two equal and opposite magnetic sublattices with zero resultant moment.

monoxide is shown in Fig. 3.5 in which nearest neighbours coupled by oxygen ions have antiparallel spins and no macroscopic magnetization results. More complicated crystal structures such as the spinel structure may produce more complicated spin structures from which a resultant magnetic moment may arise. A schematic two-dimensional representation of this case (ferrimagnetism) is shown in Fig. 3.6. The magnetic structure of antiferromagnets and ferrimagnets is therefore made up of two opposed magnetic sublattices, the *A* and *B* sublattices. In antiferromagnets the two sublattices are identical. In ferrimagnetic spinels the *A* and *B* magnetic sublattices correspond to the tetrahedral and octahedral cation sites of the structure. A resultant magnetization will generally result from the spinel $A|B_2|O_4$ in which the moment of the *B* sublattices predominates. However, even if the *B* sublattice is diluted

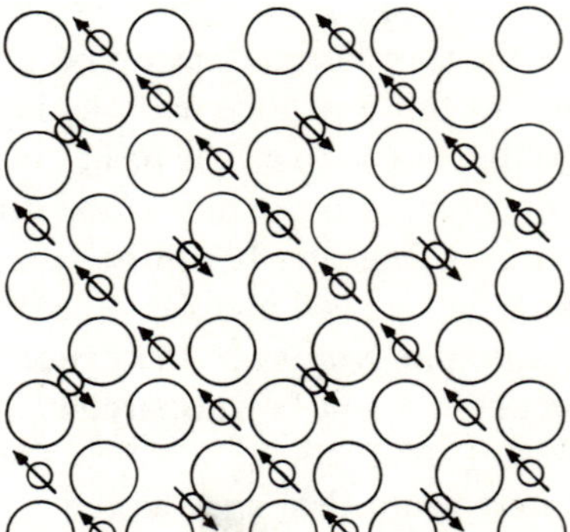

Figure 3.6 A schematic two-dimensional representation of a ferrimagnetic oxide in which cations on inequivalent sites are linked via oxygen ions to produce a spin structure composed of two opposing but unequal magnetic sublattices. Assigning the spin directions to a diagonal axis has implications about the anisotropy (section 3.4) as (isotropic) exchange would be equally satisfied with the spins orientated along any axis provided the mutually antiparallel orientations of *A* and *B* sublattices were maintained.

with non-magnetic species so that the magnetic population of the two sublattices is the same, the magnetic moments may only cancel exactly at the absolute zero of temperature. This is because in addition to the A–B intersublattice interaction there also exist weaker A–A and B–B intrasublattice interactions (section 3.3) which may be in competition with the A–B antiparallel structure. If the A–B interaction is weakened by gross dilution of one of the sublattices, the intrasublattice interaction may dominate producing antiferromagnetic order within (say) the B sublattice or some kind of triangular spin structure. 'Canted' spin structures may also arise from the combination of an antisymmetric exchange interaction with a crystal structure of low symmetry. Such a mechanism is thought to be the origin of the 'weak ferromagnetism' of haematite (section 8.1.1).

The indirect exchange (or 'superexchange') of oxides may be described quantitatively by employing the expression derived in the quantum mechanical treatment of *direct* exchange between neighbouring atoms. The exchange interaction between an atom i and its neighbours j may be described by an exchange energy, E_{ex}, of the form $E_{ex} = -2J_{ij}\mathbf{S}_i . \Sigma\mathbf{S}_j$, J, the exchange energy constant, taking a positive or negative sign to indicate a parallel or antiparallel coupling. For a pair of identical atoms this becomes $-2JS^2 \cos\theta$ (Fig. 3.7) and it can readily be seen that the energy as a function of the mutual orientation of spins provides a force (a torque) tending to align the spins parallel (if $J > 0$) or antiparallel (if $J < 0$). The spins will be perfectly parallel or antiparallel only at the absolute zero of temperature. At finite temperatures thermal vibrations will cause the mutual orientation of the spin vectors to make random excursions away from the energy minimum. At high enough

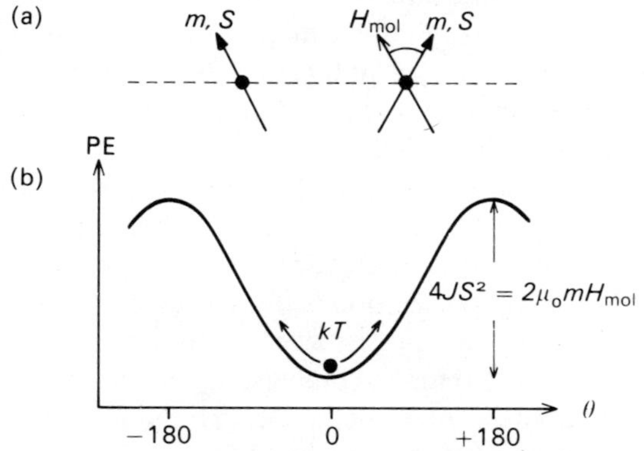

Figure 3.7 The exchange forces between a pair of magnetic atoms. (*a*) The pair of atoms with magnetic vectors lying at an angle θ with respect to one another, and the molecular field H_{mol} experienced by one due to the other. (*b*) The work done to assemble the pair at orientation θ due to the exchange interaction or the molecular field. Thermal vibrations cause excursions of the angle of mutual orientation about the energy minimum until at high enough temperature coherence is lost. The resultant magnetic moment would not be lost at a sharply defined critical point in a small system such as this. A critical point requires a system with a large number of constituents.

temperature, thermal vibrations will provide the work necessary ($4JS^2$) to rotate one spin vector against the other and the disordering effect of thermal energy (kT) will dominate over the ordering force of exchange. The temperature at which the spin structure is destroyed (it is recovered on cooling) is the Curie temperature T_C ($kT_C \sim 4JS^2$), strictly the Curie point temperature (as distinct from the asymptotic Curie temperature θ, see section 3.6). This critical point phenomenon is analogous to the melting point of a solid (other analogues are ordering in alloys, superfluidity and superconductivity). The critical point temperature for an antiferromagnet is the Néel temperature T_N.

An alternative description of the basis of the cooperative phenomenon of magnetism is found in the 'Effective Field' theories. In these theories the ordering effect of exchange is expressed as being the result of the magnetostatic interaction between a magnetic field pervading the medium, the intensity of the field itself depending on the degree of order developed in the system, and the atomic magnetic moments. An avalanche effect therefore brings about a long-range coherent order in the system (provided temperature is not too high). Consider again the simple system of two neighbouring magnetic atoms (Fig. 3.7). One atom is regarded as experiencing a 'molecular field', $\mathbf{H}_{\mathrm{mol}}$, in a direction parallel to the magnetic moment of its neighbour and equal to $\lambda_w \mathbf{m}$ where $\mathbf{m}$ is the magnetic dipole moment per atom, and λ_w is the molecular field coefficient. The potential energy of a dipole $\mathbf{m}$ in a field $\mathbf{H}$ is $-\mu_0 \mathbf{m}.\mathbf{H}$. The work done to assemble the system with mutual orientation θ is therefore $-\mu_0 m H_{\mathrm{mol}} \cos\theta$. H_{mol} is not, of course, the dipolar field due to the neighbouring atom (see section 3.4) but a device by which the cooperative phenomenon can be described. The exchange and molecular field expressions are clearly equivalent and $2JS^2 \equiv \mu_0 m H_{\mathrm{mol}}$. Thus $\lambda_w \equiv (2JS^2/\mu_0 m^2)$. The dependence of potential energy on angle is the same in either description, the height of the potential barrier opposing spin rotation being expressed in terms of either J or λ_w. Just as J takes a negative sign to describe an antiferromagnetic type of interaction, so does λ_w.

This simple system, a 'molecule' or short one-dimensional chain, will not in fact exhibit a sharp transition temperature between an ordered and a disordered state. The magnetization would decrease monotonically with ascending temperature. The persistence of spontaneous order to finite temperature requires more than one dimension and a large number of atoms. In general, approximations must be made to treat such a system in which the spin of each individual atom is coupled to its nearest neighbours which are, in turn, coupled to other atoms and so on. Further, the spins of the atoms are not classical vectors but quantum mechanical operators and the spin of each atom may occupy one of a set of eigenstates. In the Weiss molecular field approximation, all atoms are considered to be in the identical (averaged) state and each atom is influenced equally by all the atoms in the crystal. The molecular field coefficient for a ferromagnet becomes $\lambda_w =$

$(2zJS^2/N\mu_0 m^2)$ where z is the number of nearest neighbours and N is the number of atoms per unit volume. Each atom experiences a field $\lambda_w \mathbf{M}$ where $\mathbf{M}$ is the magnetization per unit volume. The spin vector is taken to be 'space quantized' taking up certain orientations to this field, each orientation having different energy, and the probability of each state being occupied at a certain temperature determined by the Boltzmann factor of statistical mechanics (for a description of this procedure, see section 3.6.1 on paramagnetic materials). Spontaneous magnetization sets in at temperatures below $T_C = 2zJS(S+1)/3k$. Above this temperature a bulk magnetization $\mathbf{M}$ exists only in the presence of an applied field $\mathbf{H}$, given by $\mathbf{M} = \chi\mathbf{H}$ (provided $\mathbf{H}$ is not too large). χ, the susceptibility, is equal to $C/(T - T_C)$ where C is a constant equal to $4N\beta^2 S(S+1)/3k$ for a 'spin only' moment (see section 3.6). Immediately below T_C, $M \propto (T_C - T)^{\frac{1}{2}}$ and at absolute zero $dM/dT = 0$.

The molecular field theory has the merits of relative simplicity while at the same time providing a qualitative description of the cooperative phenomenon and its critical point. Models aimed at providing a more quantitative fit between theory and experiment (more complex effective field theories or series expansion methods) rapidly become more complicated. Improved models predict a lower critical point than the molecular field theory. The magnetization falls more rapidly at the absolute zero of temperature with $dM/dT \propto -T^{\frac{1}{2}}$ (the 'Bloch 3/2 power law' for M) and $M \propto (T_C - T)^{1/3}$ near the Curie point. Above T_C, χ^{-1} is not linear with temperature but is curved with a smaller slope than the molecular field theory. In the high temperature limit the curve is asymptotic to a line $\chi = C/(T - \theta)$ where θ, the asymptotic Curie temperature, is higher than T_C (see section 3.6). Above the critical point temperature, the molecular field theory allows for no order whatsoever. However, above T_C the exchange forces or the effective field do provide a bias tending to produce order which will persist over a short range only and result in a 'tail' to the magnetization-temperature curve in an applied field (section 3.3). Further, and gross, differences between molecular field theory and more exact theories arise in dilute magnetic materials when the nearest neighbour sites to a magnetic atom may be occupied by a mixed population of magnetic and non-magnetic atoms which fluctuates throughout the crystal. In molecular field theory the critical point temperature simply depends linearly on concentration of magnetic atoms, i.e. each atom has an averaged magnetic coordination number. Below a critical concentration, however, the magnetic coordination number may frequently be zero, magnetic order will occur only within isolated clusters, and long-range order will not percolate throughout the crystal.

3.3 Curie temperatures of spinel oxides

The Curie point temperatures of the spinel oxide minerals are of some interest as providing a means of determining the composition of mineral phases in rocks. The Curie point is also of interest by implication in many aspects of models of magnetic behaviour in which exchange interactions play a role (e.g. domain wall energies, Chapter 4) or as providing an upper limit to the temperature range in which thermoremanance may be acquired (Chapter 5). The variation of Curie point with concentration and type of magnetic species is therefore of some importance. It may be noted that the critical temperatures as experimentally determined are not necessarily so marked that they are unambiguous (within experimental error), and conventions have to be adopted to define the Curie point (extrapolation of $(dM/dT)_{\max}$ to the temperature axis, for example). This is because the absolutely sharply defined critical point is a fiction of simple models for simple materials which accurately describe real materials only insofar as they share the simplicity of the models.

Néel adapted the classical molecular field theory to describe anti-ferromagnetism and ferrimagnetism and so obtained model expressions for the critical point temperatures in terms of molecular field coefficients, or equivalent exchange energy constants, and the numbers and types of magnetic ions occupying the sublattices. Each ion experiences a field due to the ions on both sublattices and the Curie temperature of a ferrimagnet depends on a number (at least three) of inter-(*AB*) and intra-(*AA* and *BB*) sublattice exchange constants. Neglecting the *AA* and *BB* interactions and allowing only one species of cation on each sublattice results in the simplified expression, $T_C = 2(z_A z_B)^{\frac{1}{2}} J_{AB}(\bar{A}\bar{B})^{\frac{1}{2}}/3k$, where z_A is the number of nearest *A*-sublattice neighbours to a *B* site ion, $\bar{A} = S_A(S_A + 1)$, S_A is the spin of the *A*-sublattice occupants, and J_{AB} is the intersublattice exchange constant for the particular pair of ions considered. For the spinel lattice the maximum value of z_A is 6 and z_B, 12, giving $T_C = 4\sqrt{2}J_{AB}(\bar{A}\bar{B})^{\frac{1}{2}}/k$. The tendency of molecular field theory to overestimate critical point temperatures is readily seen by comparing this last expression with that of Wojtowics (1960) who used the high temperature expansion method to obtain $T_C = 4.55 J_{AB}(\bar{A} - 0.090)^{\frac{1}{2}}(\bar{B} - 0.185)^{\frac{1}{2}}/k$. The absolute values of exchange constants inferred from measured Curie points using molecular field theory are therefore likely to be grossly underestimated (by approximately a factor of 50% from the above expressions with $S_A = S_B = \frac{1}{2}$). The relative values so obtained are, however, useful in discussing the influence of cation distribution and concentration (provided this is not too low) on the Curie temperatures of spinels. In a diluted material, z_A and z_B become, in molecular field theory, $\bar{z}_A$ and $\bar{z}_B$, the averaged values. Thus $T_C \propto (\bar{z}_A \bar{z}_B)^{\frac{1}{2}}$. The applicability of this expression has been illustrated for the case of $MgFe_2O_4$ in which the Fe^{3+} has been substituted by various non-magnetic species (see Broese van Grenou *et al.*, 1968/69) giving a range of $(Fe_A^{3+} Fe_B^{3+})^{\frac{1}{2}}$ (concentrations pfu) between about 0.3 and 1.0 and a corresponding Curie temperature range 250–650 K. Although the dependence of T_C on total concentration of magnetic species is quite strong, the dependence on cation distribution, i.e. concentrations within the sublattices, for a fixed total concentration is much less so. Taking magnesium ferrite as an example, the cation distribution may be written $Fe^{3+}_{1-x}Mg^{2+}_{x}[Fe^{3+}_{1+x}Mg^{2+}_{1-x}]O_4$ with a Curie temperature $T_C \propto (1-x^2)^{\frac{1}{2}}$. An ordered cation distribution corresponds to $x=0$ and a random distribution to $x=\frac{1}{3}$, giving a model Curie temperature difference of about 6%.

A useful adaptation to the molecular field model of ferrimagnetism has been made by Stephenson (1972) which applies to ferrites containing two species of cation. In this model the intrasublattice interactions are again neglected and each cation experiences a two-part molecular field due to the two species on the opposite sublattice. Of particular interest is the formulation applying to spinels containing Fe^{3+} and Fe^{2+} as the magnetic species (i.e. as in titanomagnetite/titanomaghemite) in which the exchange constants are determined from the measured Curie points of specific compounds. The end members of the titanomagnetite series, magnetite and ülvospinel, are used to evaluate the $Fe^{3+}-Fe^{2+}$ and $Fe^{2+}-Fe^{2+}$ exchange constants and the method therefore essentially allows the calculation of Curie points relative to the measured accepted values at the two ends of the titanomagnetite series. In simplified form, the derived expression is $T_C = 700(D-\Delta/D)^{\frac{1}{2}}$ where

$$D = [Fe_A^{3+}][Fe_B^{3+}] + 0.451\{[Fe_A^{3+}][Fe_B^{2+}] + [Fe_A^{2+}][F_B^{3+}]\} \\ + 0.0438[Fe_A^{2+}][Fe_B^{2+}]$$

and

$$\Delta = 0.0583[Fe_A^{3+}][Fe_B^{3+}][Fe_A^{2+}][Fe_B^{2+}]$$

and ionic concentrations are per formula unit. In its original form the expression also includes the factor $(8.4/a)^{13}$ where a is the cell edge in Å. This factor is mainly to take into account a presumed contraction and expansion of the Fe^{3+}–O and Fe^{2+}–O distances as cell edge varies with composition. However, the cell edge parameter, as experimentally determined, is a satistic, and on a simple 'rigid sphere' model the individual action-anion distances remain invariant, although the average value may change. Exchange constants would therefore also be invariant. To test this supposed invariance using molecular field theory and critical point temperatures is not possible because of the crudeness of molecular field theory. In the high temperature limit, however, the theory approaches exactness and the extrapolated asymptotic Curie temperature (section 3.6) may reveal a variation in exchange constant with composition in a solid solution series, which does result from actual variations in cation–anion distances.

The application of these models to the composition dependence of Curie temperature in titanomagnetites—the effect of diamagnetic substituents and development of non-stoichiometry—and the effect of cation distribution is discussed in Chapter 7. The temperature variation of spontaneous magnetization of spinel ferrites below the Curie temperature is considered in section 3.6.3.

3.4 Magnetocrystalline anisotropy

In common with all physical properties of crystalline solids, magnetism is anisotropic. This anisotropy manifests itself in many ways, those most pertinent in the context of rock magnetism being the existence of stable remanent magnetization. Without anisotropy, magnetic recording would not be a possibility, either in the natural or the man-made world. (In practice, anisotropies of origins other than crystalline may also be important in fixing a stable remanence; see sections 3.5 and 4.1). Exchange interactions produce a coherent spin structure as indicated schematically in Fig. 3.5 and 3.6. The exchange constant J has (up to now) been taken as isotropic, i.e. as long as the mutual orientation within the spin structure is preserved, the overall direction is immaterial and energy remains a minimum. The representations of the spin structures as lying in special orientations in Fig. 3.5 and 3.6, and not in some random direction with respect to the crystal structure, implies an energy minimum for the indicated orientation. To rotate the spin vectors (which remain coherently orientated) away from these preferred orientations would require work to be done. This might be provided by thermal vibrations, in which case the sample of magnetic material becomes 'superparamagnetic', i.e. the spin vectors undergo continuous random rotation in unison, (see section 4.4.4). The magnetic moments of (non-superparamagnetic) monodomain particles (section 4.3.1) will rest along the preferred direction in the absence of a field, as will the moments of the individual domains within a multidomain particle (section 4.3.2). A magnetized body will physically attempt to rotate itself to present its preferred direction ('easy' direction) to an external field and so reduce its energy. The magnetization curves (M–H curves) of a body will in general be different in different directions with M building up more rapidly with applied field in the 'easy' direction than in the 'hard' direction (e.g. figure 4.10).

Magnetocrystalline anisotropy may have its origin in a number of physical mechanisms.

3.4.1 *Dipolar anisotropy*

Consider two atomic magnetic dipoles of moment m separated by a distance r, coupled by a ferromagnetic exchange interaction and lying at an angle θ to the axis joining the atoms (Fig. 3.8). This axis is the 'unique' symmetry axis of this microscopic 'crystal' or short chain. Each atoms sits in the dipolar field, $\mathbf{H}_{dip}$, of its neighbour. The anisotropic part of the potential energy of the system is easily evaluated by imagining the right side dipole (say) having been brought from infinity and rotated from an orientation of minimum

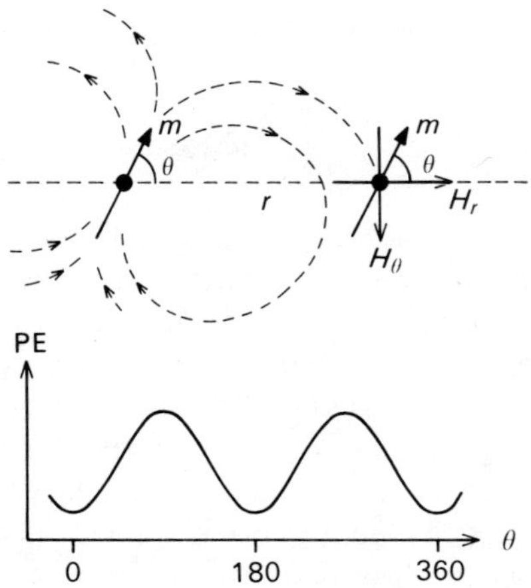

Figure 3.8 The 'pair model' for magnetocrystalline anisotropy. Two atomic magnetic dipoles coupled parallel by a ferromagnetic exchange interaction have a variable magnetostatic energy because of the orientation of each to the dipolar magnetic field of its neighbour. The energy is clearly a minimum when $\theta = 0$ as the dipoles and fields are parallel for this orientation. The energy is a maximum for $\theta = 90°$. The energy has the form $\sin^2 \theta$, characteristic of the uniaxial symmetry of this system.

energy to angle θ. The work done is then $-\mu_0 \mathbf{m}.\mathbf{H}_{dip}$. $\mathbf{H}_{dip}$ is best expressed in terms of its radial and tangential components, H_r and H_θ (Fig. 1.1) where $H_r = 2m \cos \theta / 4\pi r^3$ and $H_\theta = m \sin \theta / 4\pi r^3$. The scalar product of $\mathbf{m}$ with H_r and H_θ is easily formed, giving the potential energy as a function of orientation of the resultant magnetization of the system

$$E_{anis} = -\frac{2\mu_0 m^2}{4\pi r^3} + \frac{3\mu_0 m^2}{4\pi r^3} \cdot \sin^2 \theta$$

This is plotted in Fig. 3.8. The size and sign of the first term can be ignored and regarded simply as a base line or level. The energy has the form $E_{anis} = E_0 + K_u \sin^2 \theta$ which is a general expression for anisotropy of uniaxial symmetry (section 3.4.4). To be physically useful, the expression has to be extended to three dimensions by summing the energies of an individual dipole due to the rest of the array to give the energy of the system per pair of dipoles. The case of a one-dimensional chain and two-dimensional square lattice retain a uniaxial symmetry and the model provides a source of anisotropy. The three-dimensional cubic lattice gains no crystal anisotropy from dipolar interactions. Lattices with lower symmetry such as the corundum structure of haematite do have dipolar anisotropy, an antiferromagnetic array being described by the same expression as a ferromagnet with equivalent spin structure except for a change in sign of the coefficient of the angular function. The dipolar interaction falls off relatively slowly with distance and the total energy is a function of the shape of the body (in fact is the microscopic source of shape anisotropy, section 4.1) and the specific surface area. This mechanism may play a role in the effect of grain size on the Morin transition temperature of haematite (section 8.1.1).

3.4.2 *Single ion anisotropy*

In section 3.1 the orbital motion of the electron about the nucleus was discounted as a significant source of the magnetic moments of atoms. This is a good approximation for the transition element cations occurring commonly in minerals (predominantly Fe^{3+} and Fe^{2+}) although less so for other transition cations in oxides such as Co^{2+}. In the language of quantum mechanics, the orbital angular momentum is 'quenched' for ions in solids by the action of the crystal electric field arising from the charges of the neighbouring ions. The action of the crystal field in splitting the orbital energy levels of the $3d$ states has been described (Fig. 3.3). The five $3d$ states are degenerate in the free ion, i.e. although the states are different, distinct states, they have identical energies. When the ion is introduced into the crystal lattice, the degeneracy is reduced by the crystal field and at the same time the orbital angular momentum is 'quenched', in effect reduced or removed. The extent to which unquenched orbital angular momentum remains depends on the local symmetry of the crystal and the occupation of the states by the $3d$ electrons. When a single electron may freely transfer, i.e. without work having to be done, between degenerate states, orbital angular momentum is non-zero. Interchanging (indistinguishable) electrons between degenerate states does not correspond to a system with orbital momentum.

In octahedral symmetry, the cubic crystal field splits the $3d$ states into an upper doublet ($d\gamma$) and lower triplet ($d\varepsilon$). The unsplit $d\gamma$ states have no intrinsic orbital momentum and it is the $d\varepsilon$ states which are of interest. $3d^1(Ti^{3+})$ has one electron in the lower triplet and is therefore degenerate, the electron being free to occupy any of the three $d\varepsilon$ states. $3d^2$ (V^{3+}) is also degenerate, one electron being free to transfer to whichever of the three $d\varepsilon$ states happens to be unfilled. Similarly, $3d^6$ (Fe^{2+}) and $3d^7$ (Co^{2+}) are also degenerate and may have unquenched orbital momentum. Fe^{3+} ($3d^5$) is not degenerate even in the absence of the crystal field splitting. The degeneracy of the $d\varepsilon$ states is further removed by the field due to the trigonal symmetry of octahedral coordination. This arises from the charges of nearest neighbour anions and cations (see Fig. 2.3). The effect of cations is taken as being dominant in the energy level scheme of Fig. 3.3 in which the $d\varepsilon$ levels are further split into a lower singlet and upper doublet and a non-zero orbital angular momentum is expected for $3d^6$ and $3d^7$. The above remarks apply to the ground states of the ions in which the sum of the energies of the electrons lying in the states is the minimum possible. The electrons may be excited to higher states without violating the Pauli principle and Hund rule and recover or lose some of the orbital angular momentum by virtue of the degeneracy or otherwise of the excited system. The time-averaged orbital

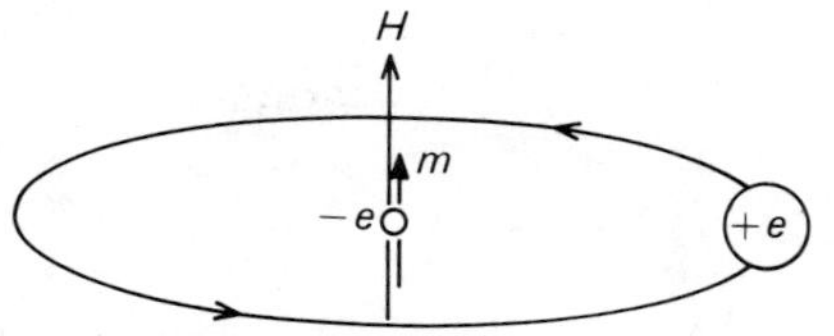

Figure 3.9 Spin-orbit coupling. In this simplified schematic picture the electron sits at the origin of a frame of reference in which the nucleus orbits the electron. The electron therefore experiences a magnetic field to which it is coupled due to its spin magnetic moment.

momentum of an ion with non-degenerate ground state will, however, be significantly smaller than one in which the ground state is degenerate. Hence the near spin-only value of Fe^{2+} ($3d^6$) of slightly in excess of 4β and the relatively large contribution of orbital angular momentum to the moment of Co^{2+}, which has 3.7β compared to the spin-only value of 3.

Although the contributions of orbital angular momentum to magnetic moments are not large, and in many cases negligible compared to the contribution from spin, the unquenched orbital angular momentum plays an important role in providing one of the connecting links between the magnetization (i.e. the spins) and the crystal lattice directions via the spin-orbit interaction. The other link is the crystal field splitting. In simple terms the spin-orbit interaction may be understood by regarding the orbiting electron as being stationary and the nucleus in motion around the electron (Fig. 3.9). The "orbiting" nucleus constitutes a current loop, because of its charge, and a magnetic field is therefore experienced by the electron at the centre of the loop. The electron is then coupled to the field by the usual $-\mu_0 \mathbf{m}.\mathbf{H}$, potential energy, $\mathbf{m}$ being related to $\mathbf{S}$, the total spin angular momentum, and $\mathbf{H}$ to $\mathbf{L}$, the orbital angular momentum. The resultant spin-orbit coupling as described by $\lambda\mathbf{L}.\mathbf{S}$ where λ, the spin-orbit coupling parameter is positive for a less than half-filled $3d$ shell and negative for a more than half full shell. Work, therefore, has to be done to rotate the spin vector (the magnetization) away from the minimum energy orientation given by the $\lambda\mathbf{L}.\mathbf{S}$ scalar product. The orientation of $\mathbf{L}$ is held more or less fixed with respect to the crystal axes by the crystal field splitting Δ. Macroscopically, work must be done in rotating the magnetization away from the preferred crystal direction and anisotropy is observed.

3.4.3 *Anisotropic exchange*

The exchange energy constant J, introduced above, may itself be anisotropic, i.e. the exchange energy may be a function not only of the mutual orientation of spins, but also of the orientation to the crystal directions of the spins en masse. Again spin-orbit coupling provides the physical mechanism. The strength of the exchange interaction depends on the degree of overlap of the

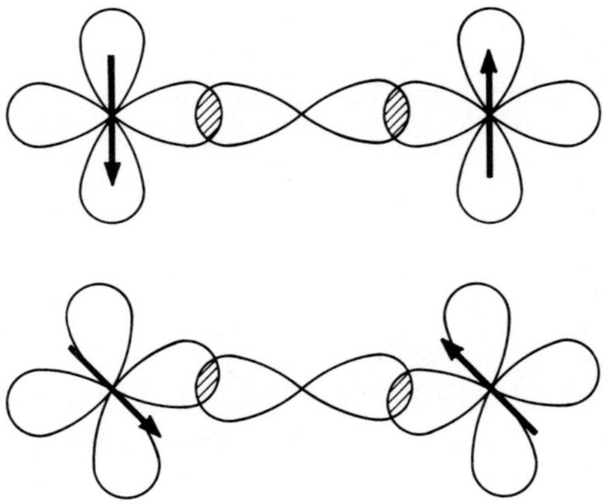

Figure 3.10 Anisotropic exchange. When the magnetization vectors (the spins) of the transition element cations are rotated (by an external field) the spin-orbit coupling exercises a torque on the orbits which changes the degree of overlap and hence the exchange energy.

orbitals of neighbouring ions. As the magnetization vector is rotated a couple is exerted on the cation orbits. They may be slightly rotated (Fig. 3.10) and the degree of overlap of the cation and anion orbitals (and hence the exchange energy) changed. The angular dependence of energy resulting from this mechanism has the same form as the classical dipole-dipole interaction energy (section 3.4.1) and can make only a small contribution to the anisotropy of crystals with cubic structure.

3.4.4 *Magnetocrystalline anisotropy constants*

For a crystal with a uniaxial symmetry, the anisotropy energy per unit volume may be written

$$E_{\text{anis}} = E_0 + K_{u_1} \sin^2 \theta + K_{u_2} \sin^4 \theta + \ldots$$

where θ is the angle between the magnetization (or the spin axis in the case of an antiferromagnet) and the unique symmetry axis. In practice it is often possible to disregard terms in K_{u_2} and above in which case the size of K_{u_1} gives the magnitude of the anisotropy, i.e. how much work must be done per unit volume to move the magnetization out of the preferred direction, and the sign of K_{u_1} tells us the preferred direction. If K_{u_1} is positive, the unique axis ($\theta = 0$) is the preferred, easy, direction. If K_{u_1} is negative, the unique plane ($\theta = 90$) is easy and the unique axis is hard. The unique plane (e.g. the basal plane of the corundum structure) itself is not isotropic. For hexagonal symmetry a term $K_3 \sin^6 \theta \cos 6\phi$ may be included where ϕ is the angle of longitude. In the case of haematite this anisotropy is difficult to detect, being generally swamped by an induced anisotropy (section 3.5) of uniaxial symmetry.

In the corundum structure, contributions to (the antiferromagnetic) K_{u_1} come from both dipole–dipole interactions and from single ion effects. The contributions have different signs, the dipolar interaction energy being a minimum for the spins lying in the basal plane ($K < 0$) and the single ion energy a minimum for the spins lying parallel to the unique trigonal axis ($K > 0$). Both effects having the same symmetry (i.e. as shown in Fig. 3.8 but one with a phase shift of 90°), the two contributions are simply additive and that with the larger size dominates. For haematite at low temperature, the single ion contribution is the greater and the spin vectors lie parallel to the trigonal axis. As temperature rises, this term falls more rapidly than the dipolar term and at the isotropic point (the Morin transition) the two terms become equal (Fig. 3.11). At higher temperatures, the dipolar term dominates and the spins now lie in the basal plane.

The expression for the angular dependence in cubic symmetry is

$$E_{\text{anis}} = E_0 + K_1\,(\alpha_1^2\alpha_2^2 + \alpha_2^2\alpha_3^2 + \alpha_3^2\alpha_1^2) + K_2\alpha_1^2\alpha_2^2\alpha_3^2 + \cdots$$

where α_1, α_2 and α_3 are the direction cosines of the magnetization vector with respect to the principal cubic axes. For magnetization confined to a principal (001) plane, $\alpha_3 = 0$ and, using the fact that $\alpha_1^2 + \alpha_2^2 + \alpha_3^2 = 1$, $\alpha_2^2 = 1 - \alpha_1^2$. Putting $\alpha_1 = \cos\phi$ where ϕ is the angle between the magnetization vector and a principal [100] axis, the energy becomes $E_{\text{anis}} = (K_1/4)\sin^2 2\phi$. This

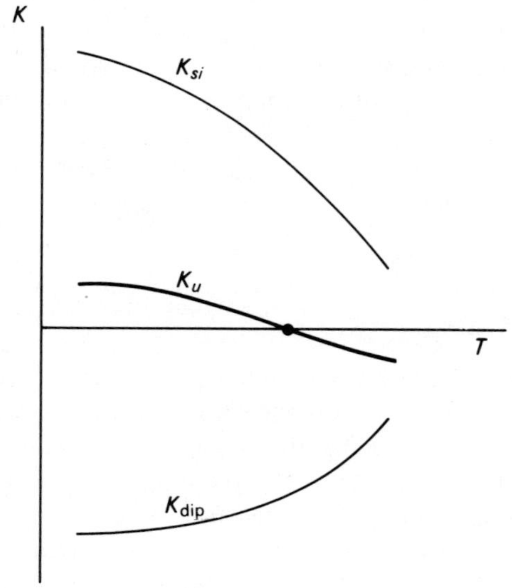

Figure 3.11 Competing (antiferromagnetic) anisotropies in the corundum structure. The positive single ion contribution, K_{si}, dominates at low temperature and the spins lie along the trigonal axis. The negative dipolar contribution, K_{dip}, falls less slowly with rising temperature. In the case of haematite the two terms cancel at the isotropic point (the Morin transition) and in the higher temperature range the spins lie in the basal plane.

provides a convenient method for the experimental determination of K_1 for single crystal materials, using a torque magnetometer. The torque experienced by a sample with magnetization confined to the (001) plane is $-(\partial E_{anis}/\partial\phi) = -(K_1/2)\sin 4\phi$. Returning to the general expression it can be seen that energies for the [100] ($\alpha_1 = 1, \alpha_2 = \alpha_3 = 0$), [110] ($\alpha_1 = \alpha_2 = 1/\sqrt{2}, \alpha_3 = 0$) and [111] ($\alpha_1 = \alpha_2 = \alpha_3 = 1/\sqrt{3}$) directions are E_0, $E_0 + K_1/4$ and $E_0 + K_1/3 + K_2/27$ respectively. It can be seen that K_1 positive corresponds to [100] and K_1 negative to [111] as easy axes respectively (provided K_2 is negligible). When K_1 is small (at or near a 'zero point', see below) the preferred direction or plane depends on the size and sign of K_2.

As the magnetic cations in a spinel oxide are coupled by exchange, the collective anisotropy of the magnetic population is simply the sum of the individual single ion contributions from all combinations of ionic species and (the two) symmetries. Some combinations (e.g. Fe^{3+} in octahedral sites) make a negative contribution to K_1, others (e.g. Fe^{3+} in tetrahedral sites) a positive one. This may result in a compensation point temperature at which K_1 is zero (similar to that shown in figure 3.11). Below this temperature the preferred direction may be (say) $\langle 100 \rangle$ and above $\langle 111 \rangle$. Whereas the specific K_1 contributions (contributions per ion) of Fe^{3+} are more or less constant and of fixed sign in octahedral and tetrahedral symmetries, that of Fe^{2+} is much more complicated. The specific magnitude and sign appear to be functions of both Fe^{2+} concentration and temperature, this latter dependence possibly accounting for the zero point in K_1 of magnetite and the titanomagnetites (section 7.1.1.4). Coming as it does from the sum of several individual contributions, each with its own individual temperature dependence, the variation of K_1 with temperature can be complicated, and cannot generally be represented by a single simple function over a wide range of temperature. In the relatively well-behaved fall in K_1 as the Curie point is approached (from below), experimental results can often be fitted by a power law of the type $K_1 \propto M_s^n(T)$ or $K_1 \propto (T_C - T)^p$.

3.5 Induced anisotropy

Anisotropy which does not arise solely from the crystal structure of the material can be regarded as an induced anisotropy. Shape anisotropy, which arises when a specimen of magnetic material or mineral lacks spherical symmetry, can thus be considered as an induced anisotropy. Discussion of this source of anisotropy will be deferred to section 4.1. Several forms of induced anisotropy are of importance in metallic alloys, partly because of their mechanical properties but mainly because the constituent atoms of (say) Fe and Ni may form superlattices with symmetries influenced by a magnetic field applied during annealing or by the process of cold-rolling ('roll anisotropy'). Magnetic annealing is also effective in inducing anisotropy in some ferrite oxides. Cooling through transition temperatures (e.g. a crystallographic transformation like the Verwey transition in magnetite, or a Curie or Néel temperature) may induce a uniaxial or even unidirectional anisotropy (exchange anisotropy, see section 5.4). The presence of a magnetic field during a change in microstructure, e.g. during precipitation, exsolution or oxidation processes, may influence the form and interrelation of the co-existing phases. Several of these processes may operate during the petrological history of a mineral assemblage, and play some part in controlling anisotropy. It seems, however, that the principal sources are shape and strain anisotropies. Strain anisotropy may have microscopic, local, importance in influencing the variation of domain wall energies with position in a crystal (section 4.2), or macroscopic influence on the overall preferred direction in a monodomain particle (section 4.3.1), or large scale regional effects on the remanence and susceptibility of the crust in areas of seismic activity or volcanism.

Magnetocrystalline anisotropy depends on the arrangement of the atoms in space (i.e. the crystal structure). As this arrangement is clearly modified in a strained crystal, a modification to the anisotropy is to be expected, the symmetry of the anisotropy being related to the symmetry of the strain. If a cubic oxide crystal is subjected to a uniaxial stress the resultant strain effects the symmetry of the oxygen anions surrounding the magnetic cations. Fig. 3.2 shows the orientation in which the cation orbitals are held by the undistorted array of anions, this orientation being connected to the macroscopic magnetization via spin-orbit coupling. In a distorted array, the cation orbitals are held in some other orientation which emerges as the symmetry axis of the anisotropy. The single-ion mechanism of strain anisotropy requires unquenched orbital momentum, just as the anisotropy of the ion in an undistorted environment does. Anisotropic ions will therefore be significant sources of strain anisotropy. A uniaxial distortion in the cubic symmetry will also allow dipole-dipole interactions to make a contribution to the anisotropy which would otherwise be zero in pure cubic symmetry. In non-cubic oxide crystals, the dipolar interaction will be modified by a change in lattice symmetry due to strain.

Just as the presence of strain modifies the preferred direction of magnetization, a crystal becomes strained on application of a magnetic field which rotates the magnetization away from the intrinsic preferred direction. Just as the lattice interacts with the spin, the spin interacts with the lattice. This is the phenomenon of linear magnetostriction. As the magnetization vector rotates, the couple exerted via the spin-orbit coupling rotates the orbitals by a small amount (as in Fig 3.10). This modifies the electrostatic repulsion between the lobes of the orbitals and the surrounding anions (in an oxide) which move to new equilibrium positions. Similarly, the energy of a dipole–dipole pair can be reduced by increasing the spacing between the dipoles. A repulsive force (dE/dr) therefore exists and the dipoles will move apart to new equilibrium positions, the strain being proportional to $\sin^2\theta$. In a system with uniaxial symmetry, the strain $(\delta l/l) = (3/2)\lambda \sin^2\theta$ where λ is the saturation magnetostriction coefficient and the system is magnetized at angle θ to the preferred direction. In the case of a cubic system

$$\begin{aligned}\delta l/l = (3/2)\lambda_{100}(\alpha_1^2\beta_1^2 + \alpha_2^2\beta_2^2 + \alpha_3^2\beta_3^2 - 1/3) \\ + 3\lambda_{111}(\alpha_1\alpha_2\beta_1\beta_2 + \alpha_2\alpha_3\beta_2\beta_3 + \alpha_3\alpha_1\beta_3\beta_1)\end{aligned}$$

where λ_{100} and λ_{111} are the saturation magnetostriction constants along the [100] and [111] directions and α_i and β_i are the direction cosines respectively of the magnetization and direction along which the magnetostriction is being measured. If, for simplicity, the magnetostriction is taken as being isotropic, $\lambda_{100} = \lambda_{111} = \lambda_s$, then $\delta l/l = \frac{3}{2}\lambda_s(\cos^2\theta - \frac{1}{3})$ where θ is the angle between the magnetization vector and the direction along which the strain is being observed. Strain anisotropy can then be expressed, in terms of the magnetostriction constant and an applied stress σ, as $E_{\text{anis}} = -\frac{3}{2}\lambda_s\sigma\cos^2\theta$, where θ is the angle between the direction of magnetization and the axis of the applied stress. The stress axis emerges as the preferred direction if $\lambda_s\sigma > 0$, i.e. the combination of a compression and negative λ_s or a tension and positive λ_s.

3.6 Categories of magnetic behaviour

Materials may be categorized magnetically on the basis of the concentration of magnetic atoms or ions and the nature of any resultant exchange interactions. Those containing no atoms with uncompensated spins e.g. H_2O, SiO_2, Mg_2SiO_4, are diamagnetic. Materials with a low concentration of magnetic species in which exchange interactions only couple isolated small clusters of atoms, and thus cannot produce long-range magnetic order, are paramagnetic. In more magnetically concentrated materials exchange interactions produce magnetic spin order at finite temperatures, the crystal structure of the compound determining whether the result is ferromagnetism or antiferromagnetism. In the latter case, given a special crystal structure, e.g. the spinel structure, ferrimagnetism may result. On a macroscopic level the various categories may be distinguished and classified by the high-temperature variation of magnetic susceptibility. In the ordered state the properties of ferromagnetic and ferrimagnetic materials are generally, of course, quite distinct from the other categories, not least in their ability to carry a remanent magnetization.

3.6.1 *Diamagnetism and paramagnetism*

The electron orbits the nucleus in the presence of the central electrostatic field. Adding also the force due to a magnetic field to the equation of motion results in a modified motion of the electron. The magnetic field (and hence magnetic moment) produced by the modified motion opposes the inducing field. A classical treatment, which produces the same result as a quantum mechanical treatment for monatomic molecules, gives for the volume susceptibility

$$\chi = -\frac{N\mu_0 e^2 Z\overline{r^2}}{6m}$$

where the atom has Z electrons with the mean square radius of orbits equal to $\overline{r^2}$. Diamagnetic volume susceptibilities are negative and small, of the order of 10^{-5} or less and vary with temperature only insofar as the geometry of the orbits is temperature-dependent. Diamagnetism is a property of all matter. The mass susceptibility of quartz is 0.57×10^{-8} m^3/kg and forsterite 0.39×10^{-8} m^3/kg.

In paramagnetic materials the magnetic atoms or ions have permanent intrinsic magnetic moments, occur in low concentrations, and the susceptibility arises from the competition between the aligning effect of the applied magnetic field and the randomizing effect of thermal vibrations. The spin vector (and magnetic moment of an ion) is space quantized, i.e. can be considered as taking up only certain orientations to the applied field. According to quantum mechanics, the total spin angular momentum of a system of electrons (such as those in the $3d$ states) is $\{S(S+1)\}^{\frac{1}{2}}\hbar$ but the values of a single component, i.e. along a specified direction, such as the direction of an applied magnetic field, are $S\hbar$, $(S-1)\hbar, \ldots, -(S-1)\hbar, -S\hbar$, i.e. the z components of the total vector which lies at discrete orientations to the specified direction (Fig. 3.12). The energies of the states due to the applied field are $\mu_0 2S\beta H, \mu_0 2(S-1)\beta H, \ldots, -\mu_0 2(S-1)\beta H, -\mu_0 2S\beta H$, the lower energy states being those with the larger components of magnetic moment in the field direction (β is the Bohr magneton). Statistical mechanics is then used to determine the average population of the states by the atoms at temperature T, using the Boltzmann factor $\exp(-E/kT)$, the probability of occupation of a state with energy E. The magnetic moment as a function of

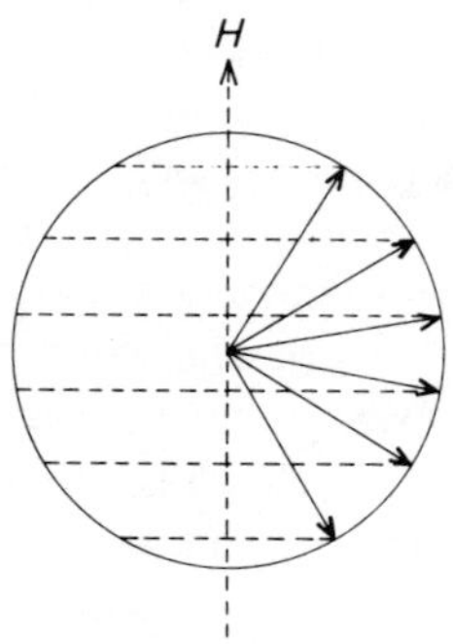

Figure 3.12 The vector model for the spin angular momentum and magnetic moment of an atom or ion, in this case $3d^5$, $S = \frac{5}{2}(Fe^{3+})$. The spin system may be in one of $(2S+1) = 6$ states, each with different energy, between $+\mu_0 5\beta H$, and $-\mu_0 5\beta H$, due to the applied field

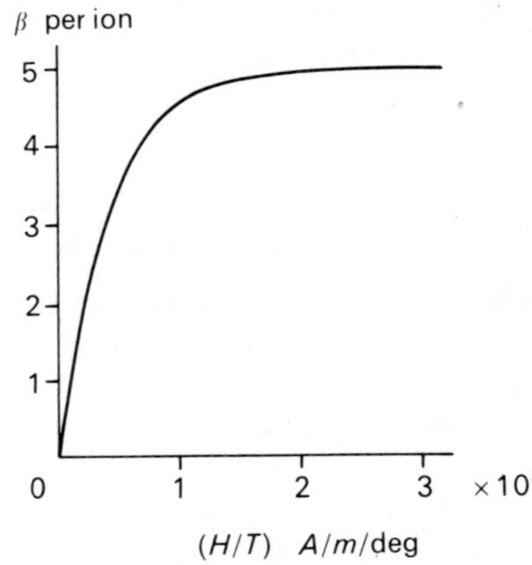

Figure 3.13 The Brillouin function describing the magnetization against field for a paramagnetic array of Fe^{3+} ions ($S = \frac{5}{2}$, magnetic moment $= 5\beta$). The magnetization is expressed as Bohr magnetons per magnetic ion and the field in units of H/T. At 300 K, the maximum field on the diagram corresponds to about 10^9 A/m, compared to the 2×10^6 A/m achievable with an electromagnet. In practice a paramagnetic material can be regarded as having a linear M–H relationship (i.e. a single valued χ).

field divided by temperature is then given by the Brillouin function (Fig. 3.13). In the limit of very large S, i.e. when there are many states and the limits imposed on the allowed orientations by the space quantization become less restrictive, this becomes the Langevin function appropriate to superparamagnetism (section 4.4.4).

The molecular field theories of ferro-, ferri- and antiferromagnetism use this Brillouin function dependence of magnetization, the field being composed of the applied field and the molecular field taken as proportional to the magnetization itself. A graphical solution is then employed to find the spontaneous magnetization. In the case of low field and/or high temperature, the Brillouin function for a paramagnetic material approximates to a linear dependence of magnetization with applied field, $M = N\mu_0 S(S+1)4\beta^2 H/3kT$ or $M = \chi H$ where $\chi = C/T$ (the Curie law, Fig. 3.14). Some effects of exchange interactions between the magnetic species may be evident, especially when they are present in higher concentrations. Expressing this effect in terms of a molecular field, $H_{mol} = \lambda_w M$, and putting $M = (C/T)(H + H_{mol})$, the susceptibility (M/H) becomes $\chi = C/(T - \theta)$ where $\theta(= C\lambda_w)$ is the asymptotic Curie temperature. θ may be positive or negative depending on the sign of the exchange interactions, and the size of θ is proportional to the size of J, the exchange energy constant, and the average number of magnetic nearest neighbours.

The mass susceptibility of a paramagnetic material (e.g. a salt of a transition metal) is typically 10^{-6} m^3/kg at room temperature.

3.6.2 *Antiferromagnetism*

The high-temperature susceptibility of an antiferromagnetic material is characterized by a large negative asymptotic Curie temperature (Fig. 3.14) indicating a high concentration of strongly interacting magnetic atoms or ions. Below the Néel temperature, T_N, magnetic order sets in (Fig. 3.5) but there is no macroscopic magnetization in the absence of an applied field. The susceptibility below the ordering temperature depends on the orientation of the applied field with respect to the spin axis. Measured parallel to the spin axis, $\chi_{\parallel}$ reflects the combined action of thermal vibrations and the applied field against the AB interaction, holding the spins of the A and B sublattices antiparallel. $\chi_{\parallel}$ has a maximum value at T_N and falls towards zero at the absolute zero of temperature. The susceptibility normal to the spin axis, $\chi_{\perp}$, results from the rotation of the spins away from antiparallel orientation and is approximately independent of temperature. $\chi_{\perp}$ is always greater than $\chi_{\parallel}$ (Fig. 3.14) and the magnetostatic energy, for a given field, is therefore lower for the perpendicular case. If the field applied in the parallel direction is large enough, the spin structure will rotate as a whole, against anisotropy forces, to become perpendicular to the applied field. The observation of field-induced 'spin flopping' provides a measure of the anisotropy in single crystals of antiferromagnetics such as haematite. In polycrystalline materials with random orientation of crystallites, $\chi = \frac{1}{3}\chi_{\parallel} + \frac{2}{3}\chi_{\perp}$. Antiferromagnetic susceptibilities, at room temperature, are generally an order of magnitude lower than typical paramagnets.

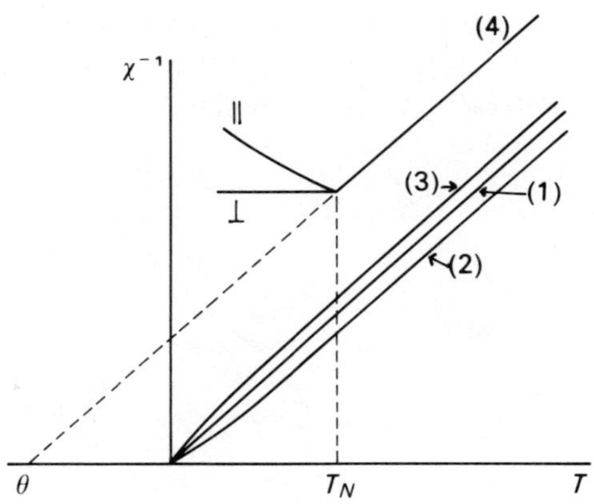

Figure 3.14 Variation of reciprocal susceptibility ($1/\chi$) with temperature T for dilute magnetic materials, and for an antiferromagnet. Line 1 corresponds to a near-ideal paramagnet, curves 2 and 3 to paramagnets in which effects due to positive and negative exchange interactions are present, and the asymptotic Curie temperature is non-zero. The antiferromagnet (4) has a large negative asymptotic Curie temperature and the susceptibility below T_N depends on orientation with respect to the spin axis.

3.6.3 *Ferrimagnetism and ferromagnetism*

Ferrimagnetics also exhibit a large, negative, asymptotic Curie temperature (Fig. 3.15) because of the large concentration of negatively interacting magnetic ions. At the Curie point, spontaneous magnetization results from the imbalance in magnetic moment between the two sublattices (Fig. 3.6), the susceptibility is therefore infinite (on a microscopic scale) and the $(1/\chi) - T$ curve is concave downwards. Below T_C the net spontaneous magnetization $M_s(T) = M_B(T) - M_A(T)$ where M_B and M_A, the sublattice magnetizations, are generally not equal because the two crystallographic sublattices contain different types and/or unequal numbers of magnetic ions. The fact that the intrasublattice AA and BB interactions may be of different sizes and signs also contributes to the inequality between the sublattice magnetizations and their different temperature dependences. The various characteristic curves of spontaneous magnetization against temperature for ferrimagnets are shown in Fig. 3.16. In a Q-type, the two sublattices both have a similar well-behaved fall in magnetization with temperature as does the resultant magnetization. P-type behaviour may arise because the dominant sublattice (the B, say) magnetization falls less quickly with ascending temperature than that of the A sublattice, because order within the B sublattice is maintained by a positive BB interaction. The same argument applies to L- and N-type, the B sublattice being progressively diluted by non-magnetic ions, so that in the case of the L-type, $M_B = M_A$ at 0°K but $M_B > M_A$ thereafter, and in the case of the N-type, $M_B < M_A$ at 0°K but $M_B > M_A$ at some finite temperature due to the more rapid fall in M_A. A compensation temperature then arises at which $M_B = M_A$. The N-type curve is of interest as a source of self-reversal of remanent magnetization (section 5.4).

Ferromagnetic materials contain magnetic atoms or ions coupled by positive exchange interactions. The susceptibility at temperatures high

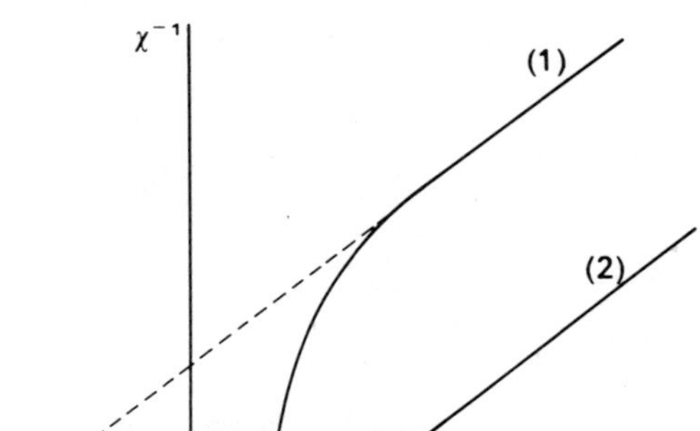

Figure 3.15 Variation of reciprocal susceptibility with temperature for (1) ferrimagnets, and (2) ferromagnets. The asymptotic Curie temperatures in the two cases are quite different because of the signs of the exchange interactions.

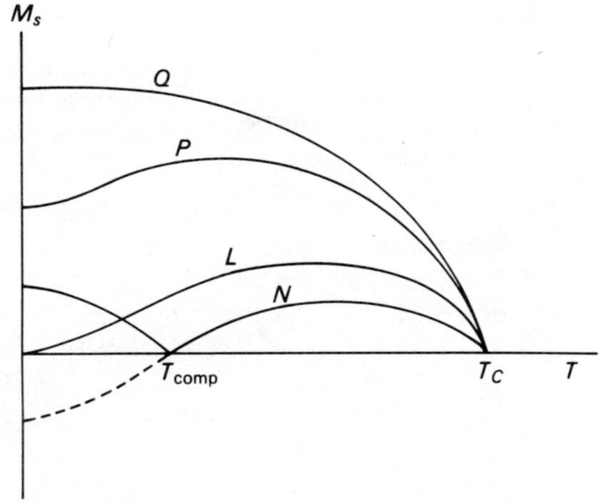

Figure 3.16 The variation of saturation magnetization with temperature for ferrimagnets. Magnetic minerals having N-type behaviour have been found but are not common.

above the Curie point produces a positive asymptotic Curie temperature somewhat higher than the Curie point temperature itself. The $(1/\chi) - T$ curve is therefore concave upwards in the region immediately above T_C (Fig. 3.16). An intermediate stage between long-range order and the disordered high temperature state occurs in this region, in which short-range order,

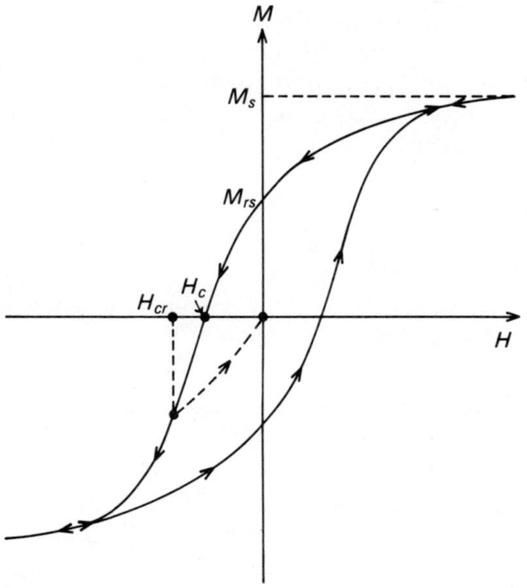

Figure 3.17 The magnetic hysteresis loop of ferrimagnets and ferromagnets. The form of the loop and numerical values of the ratios M_{rs}/M_s and H_{cr}/H_c depend on microstructure, principally grain size and shape.

or clustering, takes place. Below T_C, the $M_s - T$curve is similar to that of a Q-type ferrimagnet.

Both ferrimagnets and ferromagnets exhibit irreversibility in magnetic properties (hysteresis) below the Curie point. A hysteresis loop is shown schematically in Fig. 3.17. For a given chemical composition the form of the loop depends on microstructure—the porosity, size, shape and orientation of the magnetic grains together with their relationship to other phases—and the effects of any induced anisotropy. The characteristic features of the loop are the saturation magnetization (taken to be equal to the spontaneous magnetization when the applied field is large enough), the isothermal remanent magnetization, M_{rs}, remaining when the applied field has been removed, and the coercive force, H_c, the reverse field necessary to reduce the magnetization to zero. The coercivity of remanence, H_{cr}, is the reverse field which, when removed, leaves the specimen in the demagnetized state. The relationship between the macroscopic loop and the underlying microscopic processes is discussed in Chapter 4.

3.7 Summary

The magnetism of materials results principally from the spin angular momentum of the electron which is thus equivalent to a microscopic magnet having magnetic dipole moment 9.27×10^{-24} A m^2 (1 Bohr magneton, β). Atoms or ions of the transition series of elements have magnetic moments of up to 5β, due to uncompensated electron spins in the $3d$ states. These include Fe^{3+} ($3d^5, 5\beta$) and Fe^{2+} ($3d^6, 4\beta$ plus possibly a small contribution to magnetic moment from orbital motion of the electrons). When combined in a solid in high concentrations, magnetic atoms cooperate due to overlap of orbitals with magnetic nearest neighbours (direct exchange) or via orbital overlap with anion orbitals in oxides (indirect exchange). This results in a long-range ordered coherent spin structure. Of especial interest in the context of oxide minerals is ferrimagnetic order in which two magnetic sublattices have opposed but unequal magnetization, resulting in a net macroscopic magnetization. The magnetic order is opposed by the randomizing effect of thermal vibrations in the crystal lattice, the magnetic order disappearing at the Curie temperature.

The existence of microscopic atomic magnets and the exchange interactions are ingredients necessary to the appearance of magnetism in materials. Of equal importance in the context of many of the observed characteristics of magnetic materials is the phenomenon of magnetic anisotropy by which certain directions in the crystal structure emerge as preferred 'easy' axes. The magnetization will lie along the easy axis unless acted upon by an external field. In the absence of such anisotropy the magnetization vector would be in continuous motion and no stable magnetization would exist. In addition to anisotropy related to the way the atoms in the crystal are arranged in space (magnetocrystalline anisotropy), induced anisotropies due to the presence of strain or due to the shape of the magnetic particle may also occur.

Magnetic materials may be categorized and distinguished by the temperature variation of susceptibility. Diamagnetic materials have a negative, temperature-independent susceptibility, paramagnetic materials a positive susceptibility following the Curie law, $\chi = C/T$. For ferrimagnets, antiferromagnets, and ferromagnets, $\chi = C/(T - \theta)$ where θ is positive for ferromagnets and negative for antiferro- and ferrimagnets. Below the critical point temperatures, ferro- and ferrimagnets exhibit spontaneous magnetization and are able to carry magnetic remanence.

Selected bibliography

Broese van Grenou, A., Bongers, P.F. and Stuyts, A.L. (1968/69) Magnetism, microstructure and crystal chemistry of spinel ferrites. *Mater. Sci. Eng.* **3**, 317–392.

*Chikazumi, S. (1964) *Physics of Magnetism.* J. Wiley & Sons Inc., New York, London, Sydney.

Littlefield, T.A. and Thorley, N. (1979) *Atomic and Nuclear Physics: an introduction.* Van Nostrand Reinhold Co. New York, Cincinnati, Toronto, London, Melbourne.

*Morrish, A.H. (1965) *The Physical Principles of Magnetism.* J. Wiley & Sons Inc., New York, London, Sydney.

Stephenson, A. (1972) Spontaneous magnetization curves and Curie points of spinels containing two types of magnetic ion. *Phil. Mag.* **25**, 1213–1232.

Wojtowicz, P.J. (1960) The high temperature susceptibility of ferrimagnetic spinels. *J. Appl. Phys.* **31**, 2655–2665.

*Source material not specifically cited in text.

4 The magnetization process

4.1 The demagnetizing field—shape anisotropy

A consideration of the origin of magnetism at the atomic level is an essential step in arriving at a physical picture of magnetic materials. A satisfactory description of the macroscopic behaviour of a magnetic body is only conceivable if it is acknowledged that the body is constituted of microscopic atomic magnets having the characteristics of the inhabitants of the atomic world. In the discussions of the previous chapter, this world was considered to be infinitely large, this simplification not generally invalidating the picture so obtained. The finite size of crystals and the resultant free surface is, however, a crucial element in controlling the way in which a magnetic body, such as a mineral grain in a rock, acquires or loses magnetization in response to the presence or absence of a magnetic field, in response to rising or falling temperature and behaves, magnetically, in general. It is the free surface which results in a magnetized body producing a magnetic field and thus allows rocks to contribute to anomalies in the geomagnetic field.

A uniformly magnetized grain in the form of an ellipsoid is shown schematically in the upper part of Fig. 4.1, together with part of the magnetic field 'leaking' from such a body. Uncompensated poles occur at the surface while, within the body, the N and S poles of adjacent atomic dipoles in close proximity are effectively cancelled out. The external field arises from the surface poles and the rest of the body may be ignored, as shown in the middle diagram of Fig. 4.1. Just as in electrostatics, field lines originate and terminate at poles of opposite polarity so also in magnetostatics, as the elementary experiment with iron filings and a magnet reveals. Field lines in electrostatics radiate from poles in, or converge from, all directions and the middle diagram of the figure is clearly unbalanced in that the field lines appear on only one side of each of the array of poles. Field lines therefore also pass from N to S in the interior of the array (bottom figure). The external field is thus only half of the total picture. The internal and external fields are inseparable, interdependent and complementary parts of the total field of the body. If the external part of the field is large then so is the internal part, and vice versa. The internal part of the field is called the 'demagnetizing field', because its direction is opposed to the magnetic polarization which produces it, although it does not necessarily bring about any demagnetization. In such cases (monodomain grains, section 4.3.1) it may still play an important role as the origin of shape anisotropy. In the case of an ellipsoidal body, the

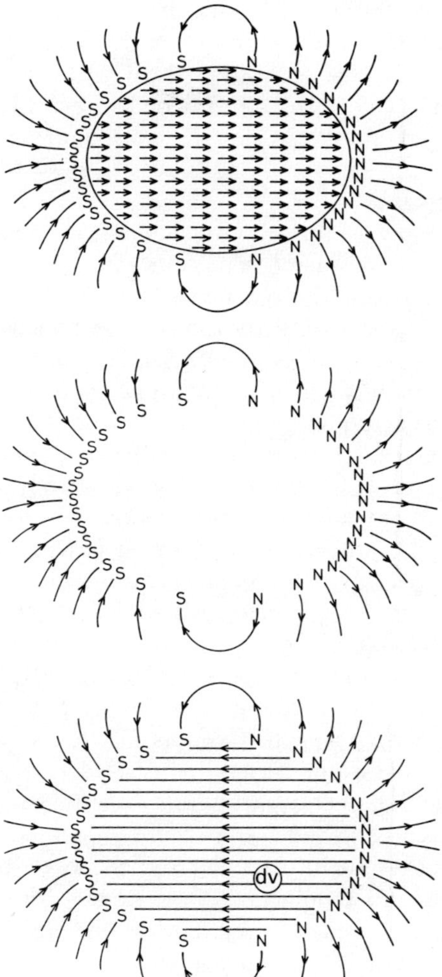

Figure 4.1 The demagnetizing field. The uppermost figure shows the internal polarization, the surface poles and the external field of a uniformly magnetized ellipsoid. The middle figure shows the production of an identical field due to an array of poles. This implies an internal field (lower figure) complementary to the external part of the total field. The elementary magnetized volume dv has a potential energy because of this internal field.

demagnetizing field, $\mathbf{H}_d$, is uniform, for uniform magnetization $\mathbf{M}$, and is given by $\mathbf{H}_d = -N\mathbf{M}$. N is the demagnetizing factor which describes the dependence of $\mathbf{H}_d$ on the geometrical arrangement of the poles, i.e. N is a geometrical factor which depends on the shape of the body. For a sphere $N = \frac{1}{3}$, for a non-spherical body N depends on the direction of magnetization—different directions producing different arrays of surface poles. An ellipsoid has $N_a + N_b + N_c = 1$ where a, b and c refer to the principal axes. An ellipsoid of revolution (about the a axis) has $N_b = N_c$ and $N_b = \frac{1}{2}(1 - N_a)$. The extreme prolate ellipsoid, an infinitely long needle, has few poles, and a large distance separating them when magnetized along the (unique) a axis, and $N_a = 0$, $N_b = \frac{1}{2}$. The opposite extreme, the oblate disk of infinite thinness, has $N_b = 0$, $N_a = 1$. The demagnetizing factor at angle θ to the unique axis for this family of bodies is $N_\theta = N_a \cos^2 \theta + N_b \sin^2 \theta$ or $N_a + (N_b - N_a) \sin^2 \theta$.

In section 3.4.1 the work done in assembling a system consisting of two magnetic dipoles was considered. The same calculation can now be made for the ellipsoidal body as a whole, the interaction field now being the uniform demagnetizing field permeating the body. The magnetostatic self-energy of the body, E_{self}, is the work done to assemble the body from elementary volumes initially at infinity. Consider the work done to place the elementary volume dv into the corresponding void in the ellipsoid of Fig. 4.1, the rest of the body having already been assembled. The magnetic moment of the element is $\mathbf{M}dv$ and the gain in potential energy of the element is $-\mu_0 \mathbf{M} dv . \mathbf{H}_d$. This operation can be repeated for each element and the work summed. As $\mathbf{M}$ and $\mathbf{H}_d$ are uniform, it is tempting to simply form $\int dv = v$. However, this counts the work between each pair of elements twice over, the work done to bring in element 1 against the field due to element 2 being mutual to the pair of elements, and not to be counted again when bringing in element 2. This means that $E_{\text{self}} = -\frac{1}{2}\mu_0 \mathbf{M} . \mathbf{H}_d v$ for a grain of volume v. This factor of $\frac{1}{2}$ is common to self-energy expressions. In section 3.4.1 the potential energy worked out was that for a pair of dipoles—the work per atomic dipole would be half the derived value. A similar factor of $\frac{1}{2}$ occurs in the binding energy of crystals—the (negative) work done per atom or ion in assembling the crystal against the attractive and repulsive force between the constituents. At a macroscopic level, the factor $\frac{1}{2}$ is shared by the energy stored in the polarized dielectric of a condenser, in a current-carrying coil and in a stretched spring. The dependence of the magnetostatic self-energy on the shape of the body is seen by substituting $\mathbf{H}_d = -N\mathbf{M}$, giving $E_{\text{self}} = \frac{1}{2}\mu_0 N M^2 v$. The further substitution for the demagnetizing factor for an ellipsoid of revolution gives $E_{\text{self}} = \frac{1}{2}\mu_0 N_a M^2 + \frac{1}{2}\mu_0 (N_b - N_a) \times M^2 \sin^2 \theta$ per unit volume, i.e. of the form $E_{\text{anis}} = E_0 + K_u \sin^2 \theta$, the uniaxial geometrical symmetry resulting in a uniaxial magnetic anisotropy. It is readily seen that K_u is positive for a prolate ellipsoid ($N_b > N_a$) and that least work is required to magnetize the ellipsoid along the long, a, axis. In fact, for a uniformly magnetized (monodomain) grain, the magnetization will lie along this axis in the absence of an external field. The oblate ellipsoid (K_u negative) requires greatest work to magnetize along the unique (but short) a axis, and least in the unique (bc) plane. The extreme cases, the needle and the disk, have $K_u = \frac{1}{4}\mu_0 M^2$ and $-\frac{1}{2}\mu_0 M^2$ respectively. For a given geometry, the size of the anisotropy depends on M^2 and this will govern how non-spherical a uniformly magnetized grain must become before shape anisotropy dominates anisotropy from other sources, and indeed it may not do so, even for the extreme geometries, depending on the magnitude of M and of the anisotropies from other sources.

The demagnetizing field is not uniform in either uniformly magnetized bodies of non-ellipsoidal shape (e.g. a cylinder) or in non-uniformly magnetized (multidomain) bodies. In either case the self-energy must strictly be determined by integrating the product of the local magnetization and internal field over the volume of the body and can no longer be determined using the demagnetizing factor. However, the self-energy may often be expressed without significant error as $\frac{1}{2}\mu_0 NM^2$, where N is an effective demagnetizing factor (section 4.3.3) and M is the averaged magnetization.

4.2 Domains and domain walls

All the energy terms relevant to the total energy of a magnetized body have now been arrived at, and the overall state of the body will be that in which the energy due to the terms is collectively minimized. To summarize, the terms are:

(i) the exchange energy $E_{ex} = -2J\mathbf{S}_1.\mathbf{S}_2$, this being a minimum for a collinear spin system;

(ii) the anisotropy energy $E_{\text{anis}} = K_u \sin^2\theta$ (for uniaxial symmetry), this being a minimum for the magnetization vector along the easy axis;

(iii) the magnetostatic self energy $E_{\text{self}} = -\frac{1}{2}\mu_0\mathbf{M}.\mathbf{H}_d$ per unit volume, this being reduced by a reduction in $\mathbf{H}_d$; and

(iv) the magnetostatic energy due to $\mathbf{H}_{\text{ext}}$, an externally applied field, $E_{ms} = -\mu_0\mathbf{M}.\mathbf{H}_{\text{ext}}$ per unit volume.

Considering $\mathbf{H}_{\text{ext}} = 0$ for the present, it can be seen that a uniformly magnetized body, such as is shown in Fig. 4.2(a) in which the magnetization vectors are parallel to each other and parallel to the easy axis, has a minimum

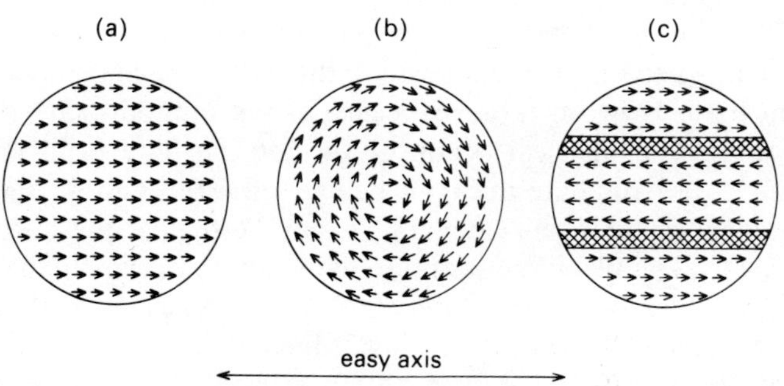

Figure 4.2 Domains and domain walls. (a) The coherent spin structure minimizes exchange and anistropy energies but maximizes magnetostatic self-energy. (b) An incoherent spin structure which removes the internal field at the cost of work done against exchange and anisotropy forces. (c) A compromise consisting of regions of coherent (domains) and incoherent (domain walls) structures.

for terms (i) and (ii) but a possibly large contribution (depending on the size of M) to term (iii). The converse holds for Fig. 4.2(*b*) in which the magnetization vectors form closed loops and the disappearance of a field leaking out of the body implies a disappearance of the internal demagnetizing field. Thus term (iii) is reduced, but contributions come from terms (i) and (ii) as the magnetization vectors are not generally parallel to either each other or the easy axis. The former spin structure can be described as coherent and the latter as incoherent. A number of styles of spin arrangement involving combinations of coherent and incoherent regions, or the presence of one or the other, have been postulated. In large enough grains, a compromise may be reached in the competition between energy terms (i) and (ii) and term (iii), and the body may contain both coherent and incoherent regions, the bulk of the volume containing a coherent spin structure, aligned with the preferred direction, separated by localized incoherent regions (Fig. 4.2(*c*)). The coherent regions are the domains and the incoherent regions the domain walls. Such a body is in the multidomain state, the larger the volume, the greater the number of domains. In small enough particles, the structure may be entirely coherent (as in Fig. 4.2(*a*)) corresponding to the monodomain state. For particle volume too large for the simple monodomain structure to represent the lowest energy state (i.e. large enough for there to be a need to reduce magnetization per unit volume, and demagnetizing field, by the onset of some form of incoherence, yet not large enough to accommodate more than a few domains), the term pseudo-single-domain has been coined. In such particles a major part of the volume may be occupied by an incoherent spin structure. Direct evidence for the multidomain state is provided by the observation of magnetic microstructure by a number of techniques, the most widely used being the colloid or 'powder pattern' technique (e.g. Soffel, 1971). Convincing evidence for the monodomain state comes from magnetic measurement. The existence of the pseudo-single-domain state as a distinct category, and confirmation of the hypothesized styles of incoherence have yet to be directly verified. The term is, however, widely used to describe the state of samples of rocks or magnetic minerals for which the measured magnetic properties are intermediate between those expected of the monodomain and multidomain states. The physical origin of the intermediate properties of the pseudo-single-domain state may not be the same throughout the range of volume associated with this category. It may be that at the lower end the particles are essentially large monodomain grains in which the large demagnetizing field initiates some degree of incoherence throughout a large part of the volume of the grain, thus reducing the effective spontaneous magnetization and assisting magnetization reversal. At the upper end of the size range the particles may be small multidomain grains in which the statistical fluctuation in the small numbers of surface domain wall nucleation points per grain results in properties different from those of larger multidomain grains.

A comprehensive review of domain patterns in oxides, and physical models describing the magnetostatics and dynamics of domain walls is provided by Rosenberg and Tanasiou (1975). Domain structures in the titanomagnetites and pyrrhotites have been studied notably by Soffel and co-workers (Soffel, 1971 and 1977; Soffel and Appel, 1982; Soffel and others, 1982) and more recently by Halgedahl and Fuller (1980 and 1981). Domain structures in haematite have been studied by, among others, Eaton and Morrish (1969). The observed patterns are variable and depend on magnetic history. In their simplest form they consist of lamellar domains (as in Fig. 4.2(*c*)) separated by more or less straight walls. Closure domains (Fig. 4.3(*a*)) may be observed where the lamellar domains meet the surface of a grain or internal free surfaces at voids. 'Spike' domains (Fig. 4.3(*b*)) have also been observed at free surfaces. Magnetization within domains lies along the preferred crystallographic directions. The angle between the magnetization in adjacent lamellar domains is 180° (i.e. the magnetization lies along the same crystal axis in both domains) and the separating walls are dubbed 180° walls. In cubic crystals with positive K_1 (i.e. easy axes at right angles along $\langle 100 \rangle$) 90° walls are possible and the closure domains of Fig. 4.3(*a*) would be made of such walls. The $\langle 111 \rangle$ axes of a cubic crystal are inclined at 70.5° so that in a material with negative K_1 the magnetizations in adjacent domains may lie at 70.5°, 109.5° and 180°. In this case the closure domains

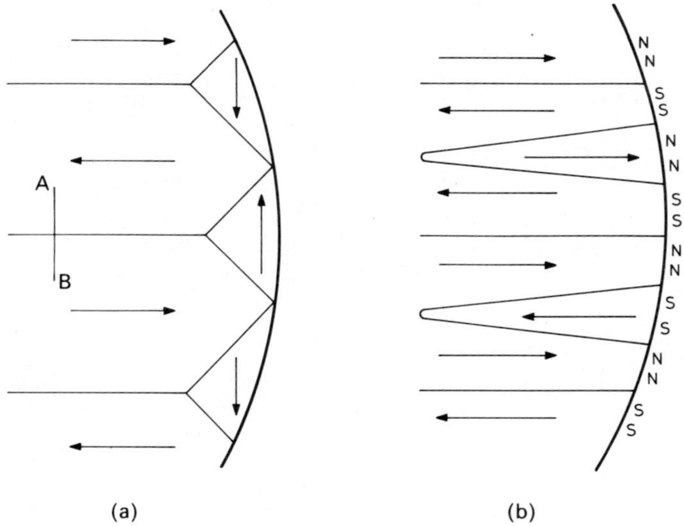

Figure 4.3 (*a*) Closure domains formed at a grain surface or internally at the free surfaces of voids. The internal lamellar domains are separated by 180° walls. Materials with positive K_1 would have the magnetization in closure domains at 90° to that in the lamellar domains, and walls between the closure and lamellar domains would be 90° walls. In the case of negative K_1, the magnetization in the closure domains would be at 71° (and 109°) to the lamellar domains, each closure domain having a 71° and 109° wall. Such closure domains will reduce the surface pole concentration. (*b*) Spike domains formed at surfaces, breaking up the surface polarity.

would be asymmetrical with the magnetizations within the closure domains not at right angles to the main domains. The role of the closure domains and spike domains, in reducing the pole concentration or breaking up the surface polarity, is clear. In some cases, e.g. after thermal demagnetization or with highly imperfect crystals, the domain pattern may be convoluted, showing no overall structure. Domain structures are three-dimensional; the observations, from which our picture of domain patterns and their behaviour is inferred, are of necessity confined to two dimensions at a free surface.

The internal spin structure of a 180° 'Bloch' wall is shown in Fig. 4.4. A single row of atoms is shown, normal to and passing through the wall, lying along the line *AB*, say, in Fig. 4.3(*a*), the wall itself consisting of many such parallel rows. The rotation through 180° on passing from one domain to the next is considered as taking place in a number of incremental steps, the spin vectors remaining parallel to the plane of the wall, i.e. as would be obtained by taking a long strip of paper with the heads of the arrows along one edge and the tails along the other and twisting one end through 180° Work has to be done against exchange and anisotropy forces, due to energy terms (i) and (ii), in constructing the wall as the spins of adjacent atoms in the row are no longer parallel to each other and the spins throughout a large part of the wall are not parallel to the preferred direction. A simple model shows that the wall has an equilibrium width resulting from the competing requirements of exchange and anisotropy. The thicker the wall, the greater the number of smaller angular increments bringing about the 180° rotation. This reduces the work done against exchange forces, in an infinitely thick wall the incremental steps becoming zero as does the work done against

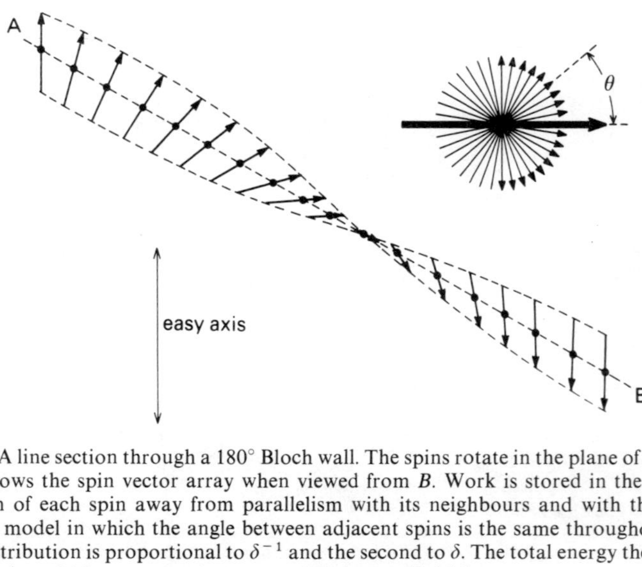

Figure 4.4 A line section through a 180° Bloch wall. The spins rotate in the plane of the wall and the inset shows the spin vector array when viewed from *B*. Work is stored in the wall due to the rotation of each spin away from parallelism with its neighbours and with the easy axis. In a simple model in which the angle between adjacent spins is the same throughout the wall, the first contribution is proportional to δ^{-1} and the second to δ. The total energy therefore has a minimum value which corresponds to an equilibrium wall thickness.

exchange. For equal angular increments, the work done against exchange would be $\propto \delta^{-1}$, where δ is the wall thickness. The thinner the wall, the fewer spins are orientated away from the easy axis. In an infinitely thin wall, i.e. made up of a single 180° increment, no work would be done against anisotropy, all spins lying in one direction or the other along the easy axis. With equal angular increments the work done against anisotropy would be $\propto \delta$. The sum of two such competing contributions ($\propto \delta^{-1}$ and δ) to the wall energy can readily be minimized, the minimum occurring in fact when the contributions are equal. In more exact models, the balance is found when the forces (torques) pulling the spins into the easy axis (making the wall thinner) and pulling the spins into parallel orientation (making the wall thicker) cancel (i.e. $dE/d\theta = 0$) at each point in the wall. It may be shown that the specific wall energy (work done per square metre), E_w, is equal to $G(A|K|)^{\frac{1}{2}}$ where $A = 2JS^2/a$, a being the inter-atomic spacing, and G is a numerical constant depending on the wall type and the crystallographic direction normal to the wall. G takes values between about 0.5 and 3 for cubic symmetry and 4 for uniaxial. E_w is typically a few mJ per square metre and the thickness of the wall of the order of 100 atomic spacings. The models which form the basis of these calculations assume an ideal crystal, and the imperfections—voids, impurities, crystal defects, local strains—which are found

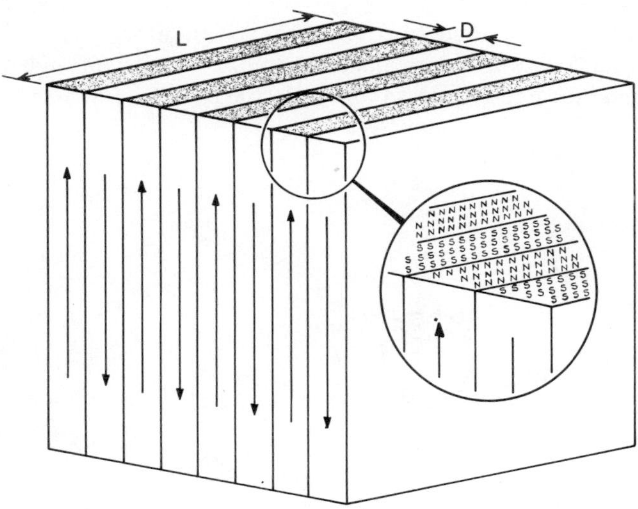

Figure 4.5 Lamellar domain structure of a rectangular plate, here shown as a cube. The smaller D, the separation between domain walls, the lower the magnetostatic self-energy. However, the smaller D, the greater the total wall area and wall energy. The magnetostatic energy per unit area is proportional to D and the wall energy per unit cross-sectional surface depends on LD^{-1}. An equilibrium value of D therefore obtains, corresponding to an energy minimum. The equilibrium domain wall spacing is proportional to $L^{\frac{1}{2}}$. In this simplified and ideal model, $D = 3.04 \times 10^{-3}(E_w L)^{\frac{1}{2}}/M_s$ where E_w is the specific wall energy. The size below which domain walls will not occur is given approximately by the condition $D = L$.

in real crystals, result in wall energies and thicknesses which are not constant but vary with the local conditions. The number of lamellar domains, or alternatively the separation of domain walls, may also be derived on an idealized model (Fig. 4.5). The magnetostatic self-energy is reduced by a greater amount the more the surface polarization is broken up by a greater number of domains. For domain wall separation D the magnetostatic energy per unit surface area is $\propto M_s^2 D$. However, the smaller the D the greater the number of domain walls and the greater the work stored in the walls. The total wall area is $\propto L^3/D$ where L is the particle dimension (the edge of a cube, say). The energy stored in the walls per unit cross sectional surface is $E_w L/D$. Again the sum of the two energies is readily minimized and the equilibrium value of D obtained. Thus $D \propto (LE_w)^{\frac{1}{2}}/M_s$, and the number of domains, $L/D \propto L^{\frac{1}{2}}$. This ideal model does not include the effects of submicroscopic microstructure on the wall energy, and the spacing and number of walls will be influenced by the fluctuations in E_w with position. Nevertheless the model values of E_w, δ and D, or at least adopted characteristic values for a given material (at a given temperature), are useful and can be regarded as mean values about which random fluctuations occur within a grain or from grain to grain.

The variation of specific wall energy with position results from any factor locally influencing the exchange energy, i.e. type of atom or variation in interatomic spacing, or the anisotropy energy, i.e. local enhancement or diminution of anisotropy due to contributions from other anisotropy sources such as strain. The wall energy depends on interatomic spacing, and internal forces tending to separate the atoms (either volume or linear magneto-

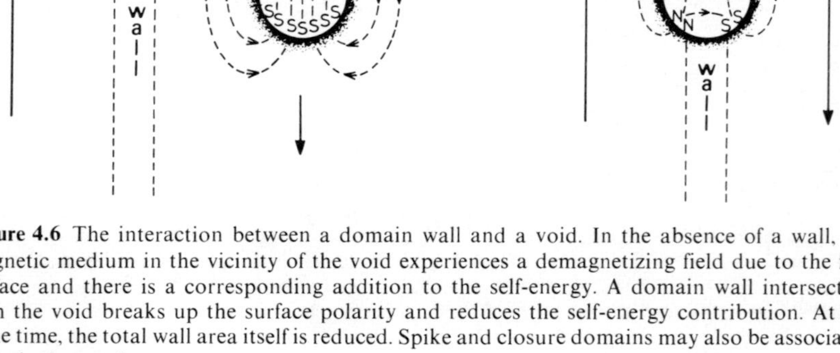

Figure 4.6 The interaction between a domain wall and a void. In the absence of a wall, the magnetic medium in the vicinity of the void experiences a demagnetizing field due to the free surface and there is a corresponding addition to the self-energy. A domain wall intersecting with the void breaks up the surface polarity and reduces the self-energy contribution. At the same time, the total wall area itself is reduced. Spike and closure domains may also be associated with the free surface.

striction) may relax depending on the local mechanical state of the crystals. Voids also play a magnetostatic role because of their free surface. The interaction between voids and a 180° wall is indicated in Fig. 4.6. A flux leakage occurs in a void located within the body of a domain, and the corresponding demagnetizing field in the material around the void results in a stored self-energy in this region. A wall passing through the void reduces the flux leakage and the demagnetizing field and self-energy. At the same time the wall area is reduced and hence the total wall energy. Further, more complicated, patterns are possible in which the void may be associated with spike domains in the absence of the 180° wall, or the 180° wall may terminate in closure domains when intersecting the void. In any event, the simple picture illustrates the source of the fluctuation of energy with domain wall position (Fig. 4.11).

As particle size is reduced, the number of domains ($\propto L^{\frac{1}{2}}$) is also reduced. When the particle size and mean domain wall separation become about the same ($L = D$) the particle may no longer contain a domain wall. The numerical constant of proportionality in the model relation between D and $L^{\frac{1}{2}}$ is $(3.04 \times 10^{-3} E_w^{\frac{1}{2}})/(\mu_0 M_s)$. Thus, for $L = D$, $L = (9.24 \times 10^{-6} E_w)/\mu_0^2 M_s^2$. In fact, the above model becomes approximate when L approaches D. An alternative, and similarly approximate, expression can be obtained by comparing the energies of a grain divided by a single domain wall through the centre (there is then no lower limit to the volume occupied by the coherent region, corresponding to D in the previous model) and the same grain uniformly magnetized in the monodomain state. This is illustrated for a spherical particle in Fig. 4.7. The energy of the monodomain particle coherently magnetized along the easy axis is the magnetostatic self-energy $\frac{1}{2}\mu_0 N M_s^2 (4\pi/3) r^3$, the demagnetizing factor N being $\frac{1}{3}$ for a sphere. The two-domain grain has reduced self-energy, the calculation of which is not simple, and for the present purpose we simply suppose that it is reduced to one half of the monodomain case. The reduction is effected at the expense of the work stored in the domain wall, i.e. $E_w . \pi r^2$. It is seen that in the monodomain case the stored energy depends on r^3 and for the two-domain case, at least partly, on r^2. For small r, $r^3 < r^2$ and the monodomain case is there-

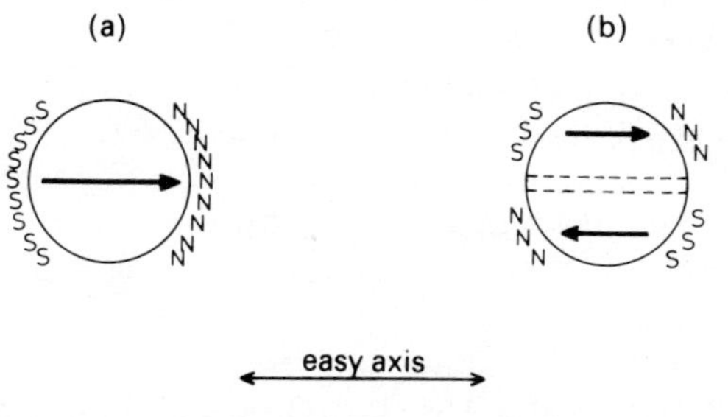

Figure 4.7 A spherical particle of radius r in (*a*) the monodomain state and (*b*) divided by a single domain wall. The self-energy of state (*a*) is $\frac{1}{2}\mu_0 N M_s^2 (4\pi/3) r^3$. In state (*b*) the self-energy may be taken as being one half of state (*a*) with additional stored energy in the wall of $E_w(\pi r^2)$.

fore favoured. For large r, $r^2 < r^3$ and the two-domain case has the lower energy. At the point of transition between the two regimes the two alternatives have the same energy. Equating the two and putting $r = r_{crit}$, we obtain $r_{crit} = (9/\mu_0)(E_w/M_s^2)$ numerically similar to the previous result.

The essential characteristic of a monodomain grain is that it cannot be demagnetized and to some extent this characteristic may be shared by small multidomain grains (pseudo-single domain grains). In the two-domain case of Fig. 4.7 the domain wall is taken to lie exactly through the centre of the spherical particle and the particle appears 'demagnetized' to an observer at a distance large compared to the particle radius. The submicroscopic microstructure of the crystal may not allow the domain wall to rest in such a position in the absence of an external field and the wall will lie in the nearest minimum energy position. The two domains will then have unequal volumes and the particle a net and irremovable magnetic moment, the size of the moment depending on the distance between the centre of the particle and the nearest stable wall position. If this distance is x, the difference in volume of the two domains is $2\pi r^2 x$ and the resultant moment $M_s^2 \pi r^2 x$. The moment per unit volume of the whole particle is then proportional to $1/r$. This moment is parallel to the magnetization in the domain having the larger volume. The equilibrium positions of the walls in 3-domain perfect crystals produce a residual moment (Craik and McIntyre, 1967; Dunlop, 1983) also, the effect reducing with increasing odd number of domains. Another source of residual irremovable magnetic moment may come from the wall itself. It can be seen from the end-on view (see Fig. 4.4) of a row of atoms in the wall, that the spin vectors have a resultant moment at right angles to the magnetization directions within the domains (up and down in the figure). The spin vectors lying at angle (see Fig. 4.4) contribute a moment per unit volume of wall of $M_s \cos\theta$ to the resultant. The average value of $\cos\theta$ on a semicircle between $-\pi/2$ and $+\pi/2$ is $2/\pi$, and for a wall volume of $\pi r^2 \delta$ the resultant total moment is $M_s(2/\pi)\pi r^2 \delta$. The moment per unit volume of the whole particle is then proportional to $1/r$. The wall of Fig. 4.4 consists of a 'right-handed' rotation of spin vectors. A left-handed rotation is equally possible producing a 'wall moment' opposite to the right-handed rotation. A single wall may be composed of regions of both right- and left-handed rotations so that the wall moment may be self-cancelling. Just as the particle divides into domains to reduce magnetostatic self-energy, so may the domain walls be divided. The opposite sense wall regions are separated by 'Bloch lines'. These contain a spin structure in which rotations occur to connect up the wall structures on either side. The work done against exchange and anisotropy per unit volume of Bloch line is higher than in the Bloch wall and the presence of the Bloch lines requires a corresponding saving in magnetostatic energy in the walls. The Bloch line itself has a resultant moment, at right angles to both the domain moments and the wall moments. The moment of a single Bloch line, if present, may be about an order of magnitude smaller than the (undivided) wall moment and is probably small enough to be disregarded as a contribution to the residual magnetic moment of many real assemblages of mineral grains containing some monodomain grains. This is not necessarily the case for moments due to a non-centred wall or due to an undivided wall depending, of course, on what fraction of an assemblage is in the monodomain state.

4.3 The magnetization process

We now introduce energy term (iv), the magnetostatic energy due to an external field, $-\mu_0\mathbf{M}.\mathbf{H}_{ext}$ per unit volume. As $\mathbf{H}_{ext}$ is increased, term (iv) becomes more important and terms (ii) and (iii) less so, i.e. the work done in having the magnetization along some non-easy direction or in having a coherent spin structure and large demagnetizing field is less than that which would be required to have the magnetization in whole or in part lying in some direction other than that of $\mathbf{H}_{ext}$. We take the three cases—the monodomain, the multidomain and particles on the monodomain/multidomain boundary or having only few domains—separately. In the present section the effects of thermal vibrations are largely disregarded, i.e. temperatures are low enough that average thermal energy, kT, is much lower than the barriers opposing magnetization change. Thermal vibrations facilitate magnetization changes, providing the energy required to cross such barriers, increasing susceptibility and decreasing coercivity (section 4.4).

4.3.1 *The monodomain case*

Consider a monodomain grain with uniaxial anisotropy in the presence of an external field applied at angle ϕ to the easy axis of the grain (Fig. 4.8). The energy of the grain is the sum of terms (ii) and (iv) if we take the spin structure as remaining coherent throughout (i.e. term (i) constant). The self-energy, term (iii), is either already included in the anisotropy term (ii) if the particle is an ellipsoid of revolution, or is a constant term if the particle is spherical with anisotropy therefore coming from some other source than shape. We are not considering the simultaneous presence of anisotropies from several sources with unrelated phases and different symmetries, although this must be a possibility in mineral grains in rocks. The energy of the grain is thus $E = K_u \sin^2\theta - \mu_0 M_s H \cos(\phi - \theta)$. The form of the energy terms and the total energy is sketched in Fig. 4.8. The anisotropy curve is of constant amplitude whereas the amplitude of the magnetostatic energy increases and the minimum in the curve deepens with increasing H. The presence of the field at angle ϕ progressively displaces the minima in the total energy profile away from the anisotropy minima. Eventually the magnetostatic term dominates and the minimum energy position and the magnetization lie along the field direction. When the field lies between $\phi = \pm 90°$ and the magnetization

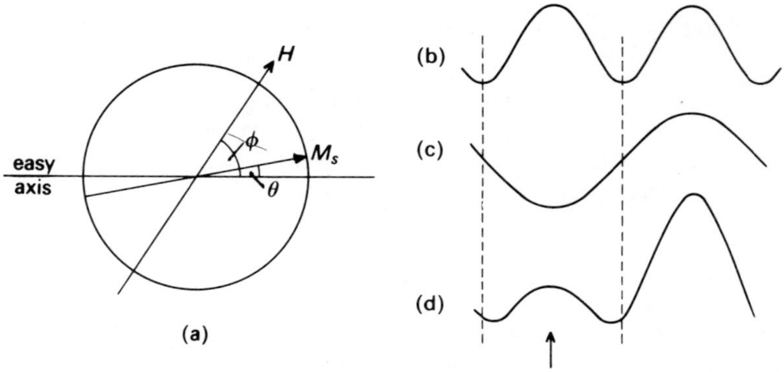

Figure 4.8 (*a*) A monodomain grain with uniaxial anisotropy energy $K_u \sin^2\theta$ and a small field applied at angle ϕ to the easy axis ($\theta = 0$), (*b*), the anisotropy energy are plotted for $\phi = 90°$ and $H = 0.6K_u/\mu_0 M_s$ and (*c*), the magnetostatic energy $-\mu_0 MH \cos(\phi - \theta)$. The lowest curve is the sum of the two energies (*d*). Both minima in the anisotropy energy are displaced towards the field direction (indicated by the arrow). An assemblage of grains has the same susceptibility regardless of the populations in the $\theta = 0°$ and $\theta = 180°$ directions. In small fields the susceptibility of a randomly orientated assemblage is $\mu_0 M_s^2/3K_u$, in the absence of thermal fluctuations.

initially lies along $\theta = 0$, it can be seen that magnetization relaxes to the original direction when the field is returned to zero. When $|\phi| > \pm 90°$, if the field is high enough for no minimum in the total energy curve to exist in the $|\phi| \leqslant \pm 90°$ zone, the magnetization relaxes into the $\theta = 180°$ position, i.e. magnetization is reversed and can only be returned to the original direction by application of a high enough field in the $|\phi| < \pm 90°$ zone.

Let us consider first low fields for which only small reversible rotations of the magnetization vector will take place. The minimum in the total energy, for a given field orientation ϕ, corresponds to $(dE/d\theta) = 0$. When the size of H is small, the excursions of M_s away from $\theta = 0$ are also small. Thus the minimum occurs at $\theta = \mu_0 M_s H \sin\phi/(2K_u + \mu_0 M_s H \cos\phi)$. The component of the magnetization along the H direction is $M_H = M_s \cos(\phi - \theta) \simeq M_s \cos\phi + M_s \sin\phi . \theta$ for small θ as before. If the susceptibility $\chi = (dM_H/dH)$ then, in the limit of small H, we obtain $\chi = \mu_0 M_s^2 \sin^2\phi/2K_u$. Thus along the easy axis ($\phi = 0, 180°$) the susceptibility is zero, as the minimum is not displaced in this case. (One might alternatively say that the susceptibility of individual grains is plus and minus infinity along the $\theta = 0$ and $\theta = 180°$ directions, as the magnetization lies along this 'easy' direction in the absence of a field, but this is not a useful idea in the context of susceptibility measurements. Thus, for an assemblage of aligned monodomain grains, the 'easy' direction is the direction of lowest susceptibility and the 'hard' direction that of highest susceptibility!) The average value of $\sin^2\phi$ for directions distributed uniformly over a sphere is 2/3. Thus the initial susceptibility (in the absence of thermal vibrations) of an assemblage of uniaxial grains with randomly orientated easy axes is $\mu_0 M_s^2/3K_u$.

For high fields in the $|\phi| > \pm 90°$ zone, the energy minimum first moves continuously and then jumps over an angular zone in which no minimum can exist before a final continuous movement into the field direction. The discontinuous movement is readily illustrated by the $\phi = 180°$ case, i.e. magnetic field opposing the magnetization. The relevant energies are sketched in Fig. 4.9. The uniaxial anisotropy energy of constant amplitude is as before.

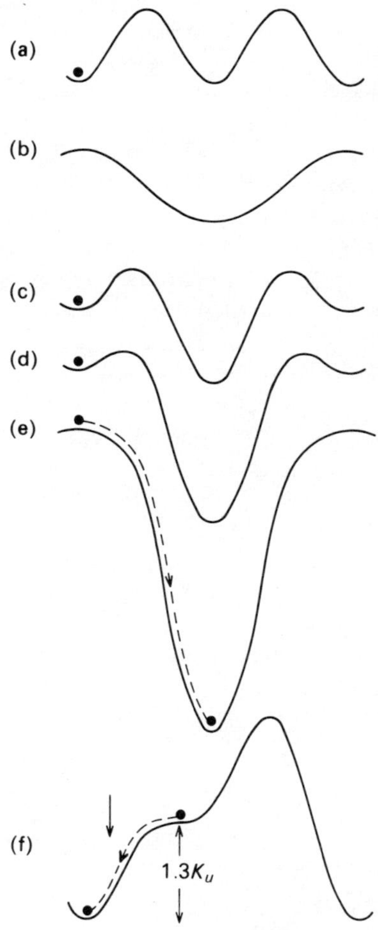

Figure 4.9 A monodomain grain with uniaxial anisotropy in the presence of higher fields. Curve (*a*) is the anisotropy energy as before and (*b*) the magnetostatic energy due to a field in the $\phi = 180°$ direction. Curves (*c*), (*d*) and (*e*) are the sum of the two energies as the magnetostatic term increases in amplitude. A magnetization lying at $\theta = 0°$ is eventually precipitated from this orientation when the field has become high enough. There is no intermediate stable position and the magnetization flips directly to the $\theta = 180°$ direction. Curve (*f*) illustrates the field applied at 45° to the $\theta = 0°$ direction. Here the two minima migrate towards the field direction (arrow) before the destruction of one minimum and the magnetization jump. The magnetization passes through the field direction during this jump and migrates back towards it on further increase in field strength.

The magnetostatic term is $\mu_0 M_s H \cos\theta$. The sequence of curves *c*, *d* and *e* corresponds to increasing H. In this case the minima in the anisotropy energy are not displaced as H increases, but one ($\theta = 180°$) is deepened and the other ($\theta = 0°$) made shallower. Eventually the rising maximum in the magnetostatic energy curve dominates the anisotropy $\theta = 0°$ minimum. The $\theta = 0°$ orientation in the total energy curve itself becomes a maximum, and the magnetization rests in an unstable equilibrium. The thermal vibrations corresponding to any non-zero temperature are enough to displace the magnetization from the equilibrium position, and the vector swings into the $\theta = 180°$ orientation, where it remains on removing the field. An increasing reverse field along $\theta = 0°$ eventually returns the magnetization to the original direction in the complementary sequence of curves. The field (H_c) required to flip the magnetization vector is found from the condition $dE/d\theta = 0$, $d^2E/d\theta^2 = 0$ at $\theta = 0$, giving $H_c = 2K_u/\mu_0 M_s$ and the hysteresis loop ($M_H - H$) is rectangular. Other field orientations have other shaped loops. The critical field for $\phi = 45°$ is $K_u/\mu_0 M_s$, and the total energy curve, in which one minimum has just become a point of inflection from which a magnetization vector will be precipitated into the other, deepened minimum, is shown in Fig. 4.9(*f*). A further increase in field translates this minimum back towards the field direction. The hysteresis loops for $\phi = (0°/180)$, $(45°/225°)$ and $(90°/270°)$ are shown in Fig. 4.10 together with the loop for a randomly orientated assemblage of grains with uniaxial anisotropy. The assemblage can be demagnetized, has a coercive force $H_c = 0.958\ K_u/\mu_0 M_s$ and a remanent magnetization after saturation (M_{rs}) of $0.5\ M_s$, when magnetization vectors are distributed uniformly over a hemisphere, about the direction of the removed field. The application of a reverse field $H_{cr} = 1.09\ H_c$, and its

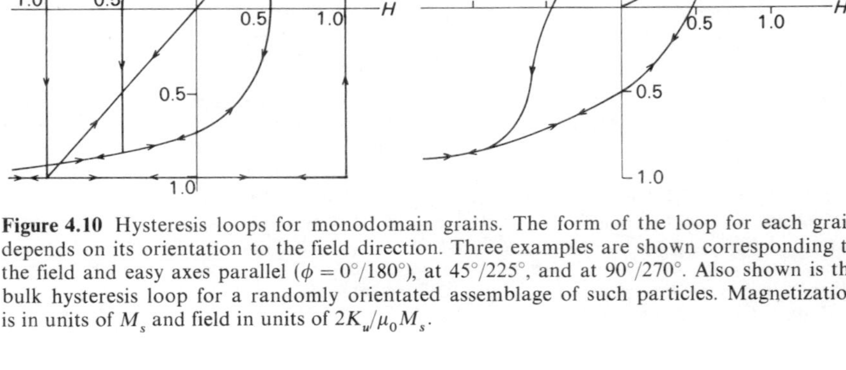

Figure 4.10 Hysteresis loops for monodomain grains. The form of the loop for each grain depends on its orientation to the field direction. Three examples are shown corresponding to the field and easy axes parallel ($\phi = 0°/180°$), at 45°/225°, and at 90°/270°. Also shown is the bulk hysteresis loop for a randomly orientated assemblage of such particles. Magnetization is in units of M_s and field in units of $2K_u/\mu_0 M_s$.

subsequent removal leaves a fraction of the grains (those in the zone $30° < \phi < 60°$, section 4.3.4) reversed. The resultants of the two complementary annular cones of magnetization vectors on opposing hemispheres exactly cancel and the assemblage is left in the demagnetized state (Fig. 4.13).

The saturated state is reached only in infinite field, which is the case for all ϕ except (0°/180°) and (90°/270°). The approach to saturation of the assemblage in high field can be described by the law $M = M_s(1 - b/H^2)$. Again, the orientation of the magnetization vectors within individual grains are given by $(dE/d\theta = 0)$ where θ and ϕ are almost the same, i.e. the angle $(\phi - \theta)$ between M and H is small and approximately $K_u \sin 2\phi/\mu_0 M_s H$. The component of the magnetization along the H direction is $M_s \cos(\phi - \theta)$ or $M_s(1 - (\phi - \theta)^2/2)$ for small $\phi - \theta$. Substituting for $(\phi - \theta)^2$ we obtain the law of approach to saturation with $b = K_u^2 \sin^2 2\phi/\mu_0^2 M_s^2$, the mean value of $\sin^2 2\phi$ for directions uniformly distributed on a hemisphere being 4/15. The work done to ascend the reversible part of the hysteresis loop (the section between M_{rs} and M_s) may also be derived from the monodomain model and is given, experimentally, by μ_0 times the area contained by the ascending/descending branch of the loop, the vertical M axis and a horizontal line $M = M_s$. For the field at orientation ϕ to the grain unique axis, the reversible work is $K_u \sin^2 \phi$ and for a randomly orientated assemblage becomes $2K_u/3$.

A uniaxial system in which the unique axis is hard ($K_u < 0$) will also display characteristic magnetization curves but without hysteresis, provided K_3 (section 3.4.4) is zero.

An anisotropy of cubic symmetry presents a rather more complicated problem. Remanence will be higher in a randomly orientated assemblage because the greater number of easy axes will allow closer clustering of the vectors about the removed field direction. For $K_1 > 0$ (cube edges preferred) $M_{rs} = 0.83\ M_s$, for $K_1 < 0$ (body diagonals) $M_r = 0.87\ M_s$, and $H_c = 0.64|K_1|/\mu_0 M_s$. The constant b in the approach to saturation law is $(8/105)(K_1^2/\mu_0^2 M_s^2)$ and the work done to ascend the reversible part of the hysteresis loop is $K_1/5$ ($K_1 > 0$) or $2|K_1|/15$ ($K_1 < 0$). The initial susceptibility is $\mu_0 M_s^2/3K_1$ ($K_1 > 0$) or $\mu_0 M_s^2/2|K_1|$ ($K_1 < 0$), for a randomly orientated assemblage.

4.3.2 *The multidomain case*

Turning now to the multidomain case, we include all four energy terms. A domain structure arises to reduce the self-energy and the variation in local domain wall energy with sub-microscopic microstructure plays an important role in the magnetization process. As the nature of the microstructure is variable and indeterminable and as a system of many, interacting, domain walls is theoretically intractable, our description of the magnetization process is qualitative and general.

Figure 4.11 gives an idealized and schematic picture of the process. We

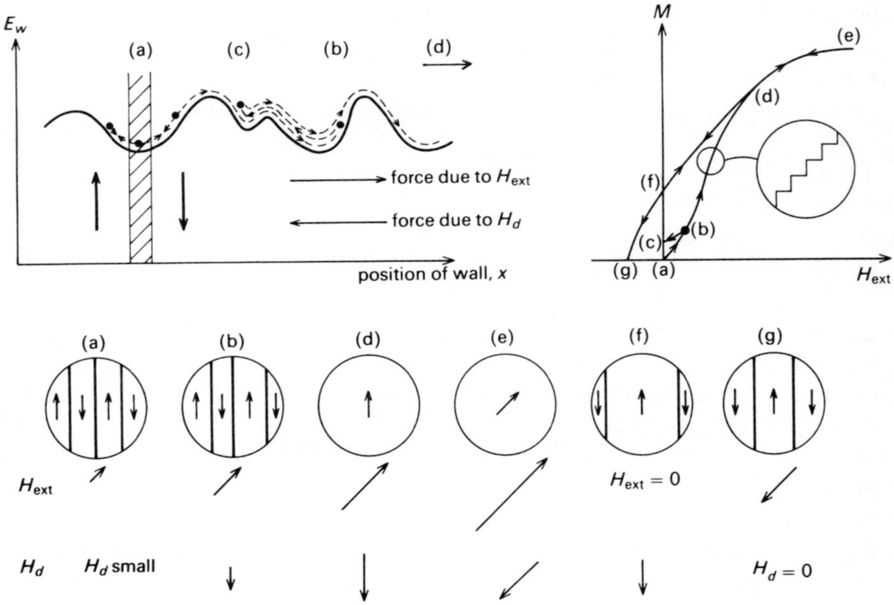

Figure 4.11 The magnetization process in a multidomain grain, a simplified and schematic picture. The sequence (*a*) to (*g*) represents the effect on the domain structure of application and removal of an applied magnetic field. The grain is not necessarily spherical, indeed some of the discussion in the text is appropriate to a rectangular prism. The process is governed by the interplay of three energy terms, the fluctuation of wall energy E_w with position being illustrated. The magnetostatic and self-energy terms are not shown but the sizes of the external field and the demagnetizing field at each stage are indicated. The resultant magnetization curve is also shown.

start with the demagnetized state (*a*) in which the 180° walls separate two sets of domains of equal volume and opposite magnetization. The domain walls lie in positions of minimum energy on the wall energy profile. On application of an external field in an arbitrary direction, those domains with magnetization having a component in the field direction have a reduced energy and those with the opposed magnetization have increased energy. Movement of the domain walls to increase the volume of the former domains and reduce the volume of the latter, opposed, domains will reduce the total energy. A force therefore exists, tending to move the walls in such a direction. If the cross-sectional area of the particle is A and the displacement of a wall from the position corresponding to the demagnetized state is x, the resultant magnetic moment is $2AxM_s$ per wall. For n walls and particle volume AL, the moment per unit volume becomes $n.2AxM_s/AL$. The potential energy of the grain in the applied field is $-\mu_0 n2AxM_sH_{ext}\cos\phi$ where ϕ is the angle between the external field and the preferred direction. This term may be added as a straight line of slope $-\mu_0 n2AM_sH_{ext}\cos\phi$ to the E_w profile, displacing the minima in the direction of increasing x. A demagnetizing field follows from the acquisition of a net magnetization and a self-energy of approximately $(1/2)\mu_0 N(n2AxM_s/AL)^2 AL$ where N is the effective demagnetizing factor in the direction parallel to the domains, $\phi = 0$. The form of this energy term is a parabola based on $x = 0$ so this term resists the translation of the energy minima towards increasing x. The location of the minima on the combined curves is obtained from the condition $(dE/dx) = 0$. This depends on the exact form of the $E_w(x)$ profile. If we suppose that near the minimum the E_w curve approximates to a parabola, $E_w = \alpha x^2$, then it is readily shown that the magnetization per unit volume in the field direction $(= 2AM_s x \cos\phi/AL$ for each of the n translated domain walls) is $(\mu_0 n^2 4A^2 M_s^2 \cos^2\phi).\, H_{ext}/(2\alpha L + 4\mu_0 n^2 A^2 N M_s^2)$. In highly imperfect crystals the wall minima will be enclosed by high, steeply rising barriers, i.e. α will be large and the susceptibility will be small. In more perfect crystals α may be small enough that $2\alpha L \ll 4\mu_0 n^2 A^2 N M_s^2$ and the susceptibility becomes simply $(\cos^2\phi/N)$. In effect, the $E_w(x)$ variation is so slight that the stable position is dictated only by the balance between the external field and self-energy terms. The component of the field normal to the domain walls, $H_{ext}\sin\phi$, also produces a resultant magnetization by slight rotation of the magnetization directions within the domains. This rotation into the direction normal to the preferred axis of magnetization is opposed by anisotropy and results in a susceptibility $(\chi_\perp)$ the same as the monodomain cases for $\phi = 90°$, discussed in the last section. The resultant magnetization in the field direction due to $\chi_\perp$ is $\chi_\perp \sin^2\phi.\, H_{ext}$ giving a total susceptibility due to the translation of domain walls, and rotation within domains, of $(1/N)\cos^2\phi + \chi_\perp \sin^2\phi$ or for a random assemblage of grains $(1/3)(1/N) + (2/3)\chi_\perp$. In practice the initial susceptibility of multidomain grains of strongly magnetic material is often simply taken to be $(1/N)$.

For small applied fields the magnetization process is reversible, the domain walls making excursions about the local energy minima corresponding to the demagnetized state. The magnetostatic energy term, a line sloping down from left to right in Fig. 4.11, is not steep enough to destabilize the local E_w minima. When the field becomes large enough so that such destabilization occurs, the domain wall translates to a new stable position ((*b*), say) at which the E_w slope is steep enough to maintain a local minimum (assisted by the self-energy term, a parabola based on $x = 0$). On removal of the external field, the demagnetizing field (i.e. the self-energy parabola) drives the wall in the reverse, decreasing x, direction (the form of the combined E_w, magnetostatic and self-energy terms are sketched, and discussed in greater detail in section 5.2.3). The walls may not return to the starting position as the driving force of the demagnetizing field (the slope of the parabola) becomes weakened as the magnetization itself is reduced. A new equilibrium position (*c*) may now exist, giving the particle a remanent magnetization. This has been acquired by irreversible domain wall movement. In progressively higher fields, the domain walls travel further to find a place at which local energy minima survive and, on removal of the fields, come to rest in positions further and further from the starting points. For a large enough field, no minima survive and the domain wall positions are governed only by the magnetostatic and self-energy terms. The domain walls will finally arrive at the surface of the grain or vanish internally when the magnetostatic energy $-\mu_0 M H_{ext} \cos\phi$ and the self-energy $(1/2)\mu_0 N M^2$ ((*d*) in Fig. 4.11) achieve a minimum with resultant magnetization equal to M_s, i.e. $H_{ext} \cos\phi = NM_s$. The magnetization process between (*a*) and (*d*) takes place in a series of 'Barkhausen jumps' during which M increases rapidly as the domain wall moves freely until becoming 'pinned' once more at a steep section in the E_w profile and requiring an increased field for further translation. The final part of the process, (*d*) to (*e*), consists of reversible rotation of magnetization, the same as the final part of the monodomain process for $\phi \neq 0°$ and 90°, and is described by the same law of approach to saturation.

Reducing the field allows the magnetization to relax back to the point (*d*) and further reduction generally allows domain walls to nucleate at the surface and travel into the body of the grain gradually reducing the magnetization and the self-energy. Removal of the external field leaves the particle (*f*) with a saturation (isothermal) remanent magnetization. A reversed field, assisted by the demagnetizing field, drives the domain walls further into the grain, eventually achieving the demagnetized state at (*g*). The application and removal of a larger reversed field, H_{cr}, returns the grain to point (*a*). In general, the domain structure for this state is not unique but will depend on how the state has been reached. (This is true also for the distribution of magnetization directions in a demagnetized assemblage of monodomain grains.) Different magnetization distributions will result from direct field demagnetization (i.e. application and removal of a reversed field), alternating field demagnetization (section 4.3.4) and thermal demagnetization (section 5.3).

Model expressions for the hysteresis parameters may be derived on the basis of a sinusoidal wall energy variation (as in section 5.2.3). For a two-domain grain containing a single wall, the hysteresis loop is a sheared rectangular loop (for $\phi = 0$) in which the magnetization change takes place not by a single jump as in the monodomain case but in ascending and descending Barkhausen staircases (Fig. 4.12) with mean slope $(1/N)$ where N is an effective demagnetizing factor (section 4.3.3). Thus $M_{rs}/H_c = 1/N$. The value of M_{rs}/M_s depends on the steepness of the E_w fluctuations (the microstructure), N (the shape) and M_s (the type of material), and will be reduced by the presence of a greater and greater number of domains. In the case of 'soft' ($H_c \sim 400\ A/m$) spherical crystals of multidomain titanomagnetite, $M_{rs}/M_s \sim 0.01$. Another model relationship believed to hold approximately for multidomain minerals in rocks is $H_{cr}/H_c = 1/(1 - N\chi)$. As χ approaches $1/N$ and $N\chi$ approaches 1, H_{cr}/H_c increases accordingly, a soft titanomagnetite sphere typically having $H_{cr}/H_c \sim 20$.

4.3.3 *The magnetization process in the monodomain/multidomain boundary region*

In multidomain grains, the fluctuating domain wall energy (principally its steepness dE_w/dx) and the demagnetizing factor strongly influence the magnetization process, whereas in monodomain grains the anisotropy is the controlling influence. For a given material, i.e. a given M_s, the multidomain state appears generally to be 'softer' having lower H_c, higher χ and lower M_{rs}/M_s. In studies in which the hysteresis parameters and susceptibility have been measured over a wide range of grain volume, embracing at either end the monodomain and multidomain states, a gradual variation in the parameters is observed between the two regimes. Although it is only possible experimentally to prepare samples of powdered material containing a range of grain size, the ranges contained within each sample are small enough that the evidence for a grain size dependence of properties is convincing. It is, however, difficult to devise experiments which can reveal the physical origin of such a dependence. Many of the hypothesized mechanisms which may operate in this grain size range involve incoherent spin structures, either as deviations from the coherent spin structure of monodomain grains (incoherent modes of magnetization reversal) or as domain walls. Other factors which may play roles are the ability of small grains to reform domain walls at grain surfaces after the removal of the walls in a saturating field, and the interaction of domain walls in small particles with the variable submicroscopic microstructure.

4.3.3.1 *Incoherent reversal modes and nucleation of domain walls.* The derivations of model expressions for the monodomain/multidomain critical sizes of section 4.2 consisted in balancing the magnetostatic self-energy of the grain and the work stored in the incoherent spin structure of the domain wall. The introduction of the energy term due to an external field removes the simple monostability of the system—both the monodomain and incoherent spin structures may become possible in the same grain. For small grains, towards the upper end of the unambiguously monodomain range, reversal of magnetization may be achieved by fields smaller than that calculated from the monodomain models of section 4.3.1. In those models the coherent spin structure is maintained during the irreversible rotation of magnetization and the grain carries the same spontaneous magnetization, M_s, at all times. The coercive forces are then $0.958\ K_u/\mu_0 M_s$ and $0.64|K_1|/\mu_0 M_s$ respectively for a random assemblage with uniaxial or cubic symmetries. The calculations of micromagnetics show that during the irreversible rotation, the spontaneous magnetization of the grain may fall below M_s by the formation of a transitory incoherent spin structure, this incoherent reversal mode being possible only for particles above a certain critical size. Below the critical size, the particles retain the coherent monodomain structure at all times. Above the critical size the particles have the coherent structure except while in the process of reversing their magnetization. This size range would then be associated with lower H_c than the monodomain models, although M_{rs}/M_s would be the same. For the hypothesized incoherent rotational mode known as 'curling', the coercive force varies as L^{-2} for grain dimension L.

The incoherent rotation modes in large monodomain grains represent a transitory incoherence. In still larger grains a semi-stable incoherence may also be produced. Although the monodomain state may be stable in zero field, in the presence of a reverse field, added to the demagnetizing field, the particle may reduce its energy by the formation of two domains separated by a domain wall initially nucleating at the surface (Craik and McIntyre, 1967). The smaller the size below the limit for the monodomain state in zero field, the larger the field needed for the nucleation of the domain wall. Once the potential formation or actual existence of a wall becomes a possibility, the submicroscopic microstructure enters the picture. Two factors arise, the fluctuating domain wall energy within the grain and the role of surface imperfections in reducing the energy barrier which a virtual domain wall must overcome on nucleating at the extremities of the grain. However, in the high applied fields appropriate to wall nucleation in this size range, the microstructure is probably unimportant. It is not the fluctuations in E_w and their steepness which matter, but the absolute size of E_w (times the wall area). The stable wall position will be governed by the external field. Removal of the field will allow a return to the monodomain state. Once the reverse field has pushed the wall far enough into the grain that the combined magnetostatic self-energy and energy due to the external field become lower than the total wall energy, the wall completes its journey through the grain in a single movement. The grain is again monodomain and remains so on removal of the reversed field (but with reversed moment). This incoherence is stable only during the presence of a reverse field in a certain size range. It may be conceivable that an internal wall position in which there is a gross reduction in E_w succeeds in trapping the wall and stabilizes the incoherent structure provided the external field is not taken above the value needed for reversal. Thus although H_c may be reduced below monodomain model values and a fraction of the grains may be conceivably be demagnetizable, the remanence from a saturating field should equal the model monodomain value.

Consider now even larger grains, above the limit for the monodomain state in zero field. On removal of a saturating field, the applied field required for nucleation of a domain wall may be zero, i.e. a domain wall nucleates spontaneously and migrates into the grain under the influence of the demagnetizing field alone. The microstructure now assumes importance. The effect of the E_w fluctuations will be considered below (section 4.3.3.2). Domain walls are believed to nucleate at surface imperfections where local wall energy is low or local self-energy is high. In effect, a potential barrier (the work done to construct the wall) obstructs the translation of the virtual domain wall from outside the grain through the surface to the interior and this barrier is lowered by surface imperfections. Halgedahl and Fuller (1980) argue that in small grains the probability of occurrence of a surface imperfection which significantly reduces the barrier to nucleation is small. In this case the demagnetizing field alone may be unable to initiate the self-demagnetization process from the saturated state and a reverse field is necessary to assist nucleation. In large grains the probability of occurrence of suitable surface imperfections should be high. This provides a mechanism for a particle size dependence M_{rs}/M_s as, in an assemblage of grains of a particular size, the fraction which remains in the monodomain state (and the overall M_{rs}/M_s) will fall as grain size increases. Applying the statistical distribution appropriate to a small number of random events, the Halgedahl and Fuller model predicts that M_{rs}/M_s will be $(1/2)\exp\{-(L/L_{\text{crit}})^{\frac{1}{2}}+1\}$ where the cubic particle dimension is L, L_{crit} is the monodomain/multidomain transition size, and uniaxial symmetry in the monodomain state ($M_{rs}/M_s = \frac{1}{2}$) is assumed. This mechanism also provides a dependence of H_c on particle size if the nucleation field is large enough that the internal E_w fluctuations become unimportant. The domain wall, once nucleated, will pass through the particle in a single jump and H_c will equal the nucleation field.

4.3.3.2 *The interaction of domain walls with internal microstructure.* We consider now the influence of the internal microstructure on the magnetization process in small grains. The fluctuating domain wall energy may be represented by the sinusoidal variation $-\Delta E_w A \cos kx$, where A is the wall area, x is the wall displacement, k has the dimensions of $(\text{length})^{-1}$ and ΔE_w is the amplitude (see sketch, section 5.3). We consider a two-domain grain with the field applied along the direction of magnetization within the domains. For the moment we will ignore the self-energy, i.e. we are considering an infinitely long needle-shaped grain or a particle in the form of an annulus. This assumption does not affect the derived model coercive force for an ellipsoidal or other shaped grain as, at $H = H_c$, the grain is demagnetized and the self-energy is zero or small. The magnetostatic energy due to the applied field is $-\mu_0 M_s 2AxH$. When the slope of this line becomes steep enough, no minima in the total energy curve survive and the domain wall translates to the edge of the grain and vanishes in a single jump. On removal of the external field, the particle remains in the saturated state as there is no demagnetizing field to nucleate a wall and drive it into the interior. The application of a reverse field providing a steep enough slope to the magnetostatic energy term is required to drive the nucleated wall through the grain. Again, once one E_w potential wall is removed so are they all, the wall passes straight through the grain and the magnetization reverses. The needle-shaped or annular two-domain particle has a rectangular hysteresis loop. The total energy $E = -\mu_0 M_s 2AxH - \Delta E_w A \cos kx$, and $dE/dx = -\mu_0 M_s 2AH + \Delta E_w Ak \sin kx$. dE/dx cannot be equal to zero (i.e. the minima are destroyed) when the first (negative) term is bigger than the maximum value (when $\sin kx = +1$) of the second (positive) term. Equating the two terms so that the minima are on the point of disappearing, $H = H_c = (\Delta E_w k)/(2\mu_0 M_s)$. Thus the coercive force depends only on the steepness of the E_w fluctuation and, as the two energy terms scale with A in the same way, the particle size vanishes. This is so only if ΔE_w itself is independent of particle size. In the Stacey and Wise model (see Stacey and Banerjee, 1974), fluctuations in E_w depend on the fluctuations in the number of defects contained within the volume of the wall as the wall translates through the grain. In effect, the wall is sampling the microstructure as it moves through the grain. If the material has a mean number of defects per unit volume of $\bar{n}$, the average number in a wall of area A and width δ is $\bar{n}A\delta$. If the defects are distributed at random, the fluctuation in the sample collected by the wall will be $(\bar{n}A\delta)^{\frac{1}{2}}$ according to the statistics of sampling random events. If $\Delta E_w A$, the fluctuation in total wall energy, is proportional to $(\bar{n}A\delta)^{\frac{1}{2}}$ then the two energy terms no longer scale together and $H_c \propto A^{-\frac{1}{2}}$ or L^{-1}, i.e. coercive force falls as particle size increases.

M_{rs} will also show a variation with grain size according to this model, the

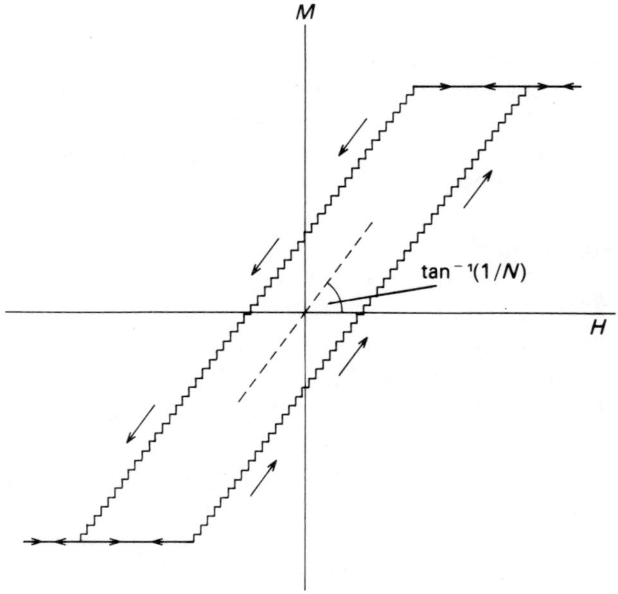

Figure 4.12 The model hysteresis loop of a two-domain particle, the field applied along the preferred direction of magnetization ($\phi = 0°/180°$). This is a sheared rectangular loop with slope $1/N$, where N is the effective demagnetizing factor. Thus $M_{rs} = H_c/N$. The ratio M_{rs}/M_s depends on H_c, and M_s all of which are independent of each other.

self-energy term now having to be considered. On the assumption that the self-energy term may be written $(\frac{1}{2})\mu_0 N(M_s 2Ax)^2(AL)^{-1}$ where N is an effective demagnetizing factor, the hysteresis loop of Fig. 4.12 is found. Dunlop (1983) has shown that for a two-domain cube the effective demagnetizing factor is constant to within about 20% and is approximately 1/6. Taking $M_{rs} \approx H_c/N$, M_{rs}/M_s will then follow the same model dependence on size as H_c, i.e. $\propto L^{-1}$.

4.3.3.3 *Irremovable moments.* Monodomain grains have irremovable moments and a magnetized assemblage has a high remanence. Because of the belief that the concentrations of monodomain grains in rocks were generally too small to account for the observed intensities of natural remanence, mechanisms have been sought by which multidomain grains could have monodomain-like properties. In two-domain grains an irremovable moment parallel to the wall may result from the absence of an E_w minimum for a centred wall, and a moment normal to the wall may come from the incoherent spin structure of the wall itself. A three-domain cubic grain without internal E_w fluctuations may have a moment because the total magnetostatic energy (now including interactions between the two domain walls) is minimized, in the absence of a field, when the central domain separating the walls is larger than (but not twice as large as) the outer domains (Craik and McIntyre, 1967; Dunlop, 1983). Detailed models for the magnetization process of these hypothetical moments have yet to be worked out. The coercive force in the first case would presumably be the same as the two-domain model of section 4.3.3.2; in the second case, H_c may be the nucleation field of a Bloch line and, in the third case, the saturating field needed to remove the domain structure entirely (E_w fluctuations being zero). In the case of the three-domain cube $M_{rs}/M_s = 0.125$ (Dunlop, 1983).

4.3.4 *Alternating field demagnetization ('fixed switching field' model)*

Alternating field (AF) demagnetization is a commonly applied operation in technology and in palaeomagnetism and rock magnetism investigations. An unwanted magnetization can be removed by randomizing the moments of an assemblage of monodomain grains or randomizing domain directions.

The fields required to bring about the irreversible rotation of magnetization of a monodomain grain with uniaxial anisotropy were discussed in section 4.3.1. The magnitude of the field required to bring about magnetization reversal depends upon the orientation of the field with respect to the unique axis, the hysteresis loops of the individual grains being shown in Fig. 4.10. For $\phi = 0/180°$ the required field is $2K_u/\mu_0 M_s$. Although the coercive force of the $\phi = 90/270°$ case is zero, the field required to uncouple the magnetization from a particular direction along the unique axis is also $2K_u/\mu_0 M_s$. The field direction for which the magnetization may be reversed most easily

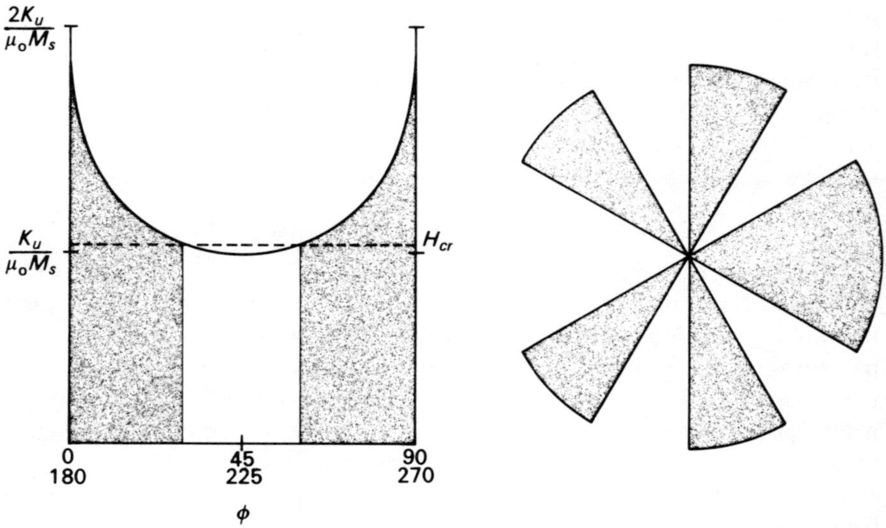

Figure 4.13 The critical field required to irreversibly rotate the magnetization vector of a monodomain grain with uniaxial anisotropy, the field applied at angle ϕ to the unique axis (c.f. Fig. 4.9 and 4.10). The field has a minimum value at $\phi = 45°$ and rises to maximum values at $\phi = 0$ and 90°. Also shown is the distribution of magnetization vector directions for an assemblage of randomly orientated identical monodomain grains first exposed to a saturating field in one direction then to a reverse field H_{cr} which leaves the assemblage with no net magnetization.

is $\phi = 45/135°$ (see also Fig. 4.9(f)). It can be shown, from the condition $dE/d\theta = 0$ and $d^2E/d\theta^2 = 0$ that the magnitude of the field, applied at angle ϕ, required to flip the magnetization vector is $(2K_u/\mu_0 M_s)(1 - w^2 + w^4)^{\frac{1}{2}}/(1 + w^2)$ where $w = (\tan \phi)^{\frac{1}{3}}$. This function, which is plotted in Fig. 4.13, is symmetrical about $\phi = 45/135°$. The effect of application of a reverse field on the remanent state of a randomly orientated assemblage of identical grains, which have been exposed to a saturating field, may be understood by reference to this figure. The field corresponds to a horizontal line in the figure, the line ascending as the field increases and the assemblage presenting all orientations ϕ to the field. When the reverse field reaches $K_u/\mu_0 M_s$ the line intersects the curve and the particle magnetizations with ϕ near 45° start to flip. (The coercive force H_c of the assemblage has already been reached before this stage, due to the reversible rotation of particles near the $\phi = 90°$ direction (Fig. 4.10); however, removal of a reverse field equal to H_c leaves the remanent state of the assemblage unchanged.) Removal of the reverse field ($> K_u/\mu_0 M_s$) leaves an annular cone of magnetization symmetrical about $\phi = 45°$ in the reverse hemisphere and a corresponding void in the initially filled 'normal' hemisphere (Fig. 4.13). The initial saturation remanence comes from the contribution $M_s \cos \phi$ for each grain, and, the average value of $\cos \phi$ over a hemisphere being 0.5, $M_{rs} = 0.5\, M_s$. The moment of the reversed annular cone is again $M_s \cos \phi$, ϕ lying between $45 + \varepsilon$ (say) and $45 - \varepsilon$. The average value of $\cos\phi$ in such an annular cone is $(\frac{1}{2})\sin 2\varepsilon$ and the reversed magnetization $(M_s/2)\sin 2\varepsilon$. The reverse field, H_{cr}, required to leave the assemblage in the demagnetized state produces a reversed magnetization equal to the residual normal magnetization, i.e. 0.25 M_s which corresponds to $\varepsilon = 15°$. The reversed annular cone lies between 30° and 60°. The horizontal line intersecting the critical field curve at $\phi = 30°$ and 60° lies at $H_{cr} = 1.048\ K_u/\mu_0 M_s$.

Turning now to AF demagnetization, this corresponds to application and removal of fields alternately in the opposite hemispheres, the amplitudes of the fields decreasing with each successive application. In practice the field will have, ideally, a decaying sinusoidal dependence on time and, in the palaeomagnetic context, the specimen may be 'tumbled' in the field. Consider the application of an alternating field with initial peak value $H_0 > 2K_u/\mu_0 M_s$ which flips the magnetization of all grain magnetizations in each cycle. As the amplitude of the field decays, the first peak value less than $2K_u/\mu_0 M_s$ is H_1 (say) and this leaves all particles in the hollow annular cone having ϕ between 0° and ϕ_1 (say) and 90° and $90 - \phi_1$ in the hemisphere corresponding to the previous ('normal') cycle (Fig. 4.14). The next peak value, H_2, in the opposite hemisphere leaves a reversed annulus between ϕ_1 and ϕ_2; H_3 leaves a normal annulus between ϕ_2 and ϕ_3, and so on. The final demagnetized state is reached as the field falls below $K_u/\mu_0 M_s$ and the actual distribution of directions depends on the decay per cycle of the alternating field, a perfect randomization requiring an infinitely slowly decaying field (see also Fig. 6.2).

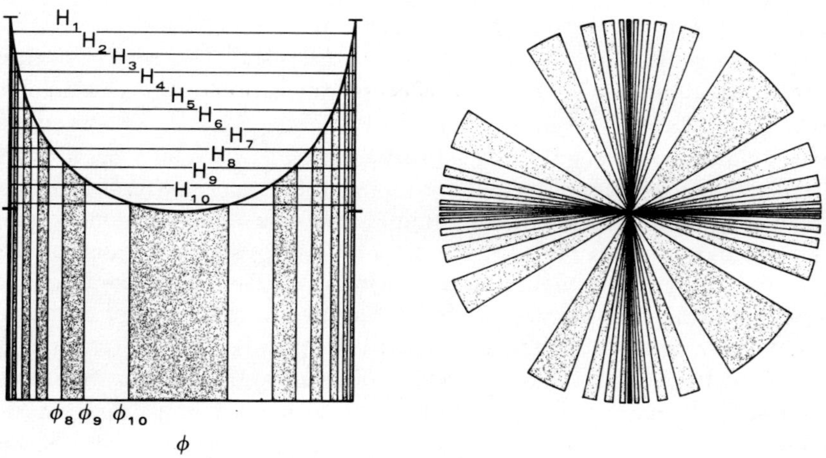

Figure 4.14 Alternating field demagnetization of an assemblage of monodomain grains. In this schematic example the peak initial value of the field before decay, H_0, is $> 2K_u/\mu_0 M_s$ and the subsequent maxima in alternating cycles during decay are H_1, H_2, etc. Also shown is the resultant distribution of the magnetization vectors in the 'normal' and 'reverse' hemisphere after the field has fallen to below $K_u/\mu_0 M_s$. A more random distribution results from a model in which thermal agitations are also included (Fig. 6.2).

A decaying alternating field with peak initial amplitude less than $K_u/\mu_0 M_s$ does not disturb whatever net magnetization the distributed grain vectors possess on this fixed switching field model. A field with initial amplitude between $K_u/\mu_0 M_s$ and $2K_u/\mu_0 M_s$ randomizes the vectors of grains located within the appropriate annular cone lying about 45°, and leaves the remainder undisturbed. Thus to ensure the removal of some kind of remanent magnetization, imprinted on the assemblage in an arbitrary direction, by application of a decaying AF field along a single direction, a field over twice as large as the coercive force of the hysteresis loop or the coercive force of remanence is required.

The demagnetization of an assemblage of randomly orientated multidomain grains may be described in a similar manner, in this case the grains with domain directions normal to the field being the first to become fixed in direction and those parallel to the field direction, the last. Again a field high compared to the coercive force of the assemblage is required. In the case of strongly magnetic materials, a peak field of the order of the saturating field is required to overcome the effects of the demagnetizing field. Just as the expected distribution of directions are different in a monodomain assemblage which has been AF and thermally demagnetized (section 5.3.) so the domain patterns within individual demagnetized multidomain grains depend on the method of demagnetization. Halgedahl and Fuller (1981) report simple linear domain patterns in AF demagnetized pyrrhotite, but undulating wall patterns in the same material after thermal demagnetization.

Resistance to AF demagnetization is reduced and directions more effectively randomized when 'thermal fluctuations' are included in the model (section 6.2.1)

4.3.5 *Rotational hysteresis*

Reversible and irreversible magnetization changes may also taken place in a field of constant amplitude but changing direction. The effect of rotating a field in a plane containing the unique axis of a uniaxial monodomain grain can be seen from Figs. 4.8, 4.9 and 4.14. In Fig. 4.8, corresponding to a small applied field, the magnetostatic energy curve would move continuously sideways as the field rotates. Both minima in the total energy curve are preserved (for small enough field) but oscillate up and down and from side to side as the field rotates. No irreversible magnetization changes take place until the field reaches $K_u/\mu_0 M_s$ (Fig. 4.14). When the angle between the unique axis and the field reaches 135°, one minimum in the total energy curve becomes a point of inflection (Fig. 4.9(*f*)) and the magnetization vector makes an irreversible jump. This occurs a second time in the $\phi = 0°$ to 360° cycle when $\phi = 315°$. The irreversible loss of energy is the difference between the energy of the grain before and after the jump of the magnetization vector. This may be read off Fig. 4.9(*f*) as 1.3 K_u J/m^3, giving a loss per

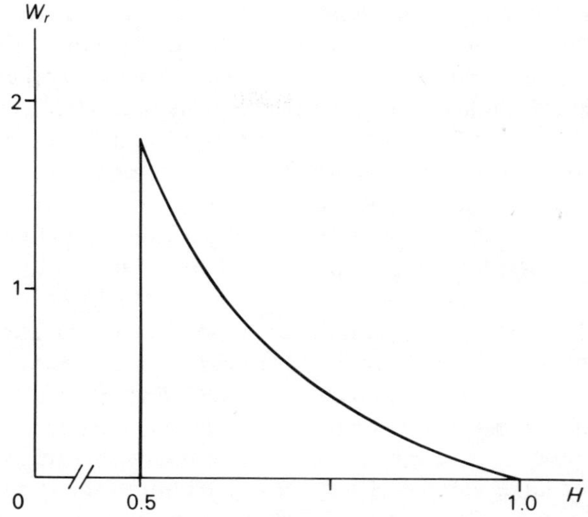

Figure 4.15 The rotational hysteresis loss per unit volume per cycle (W_r) as a function of the applied field for a random array of uniaxial monodomain grains. W_r is in units of K_u and H in units of $2K_u/\mu_0 M_s$.

cycle of 2.6 K_u. Rotating fields higher than $K_u/\mu_0 M_s$ will be accompanied by irreversible flips occurring at angle given by Fig. 4.14. This figure does not tell us where the magnetization lies before and after the flip nor what the energy loss will be. This must be found by plotting curves of the type in Fig. 4.9 or by other methods. In fact the loss per cycle decreases as H increases. When $H > 2K_u/\mu_0 M_s$ only one minimum survives in the total energy curve (Fig. 4.9(*e*)) for all orientations of the field, no irreversible flips take place and the loss becomes zero. The loss per unit volume per cycle is the same for a random planar array (with rotation in the plane) of distributed unique axes. The loss for random axes distributed on a sphere is shown in Fig. 4.15, the peak loss per unit volume being 2/3 of that for a single particle (with unique axis in the plane of rotation). A random array of particles with cubic anisotropy has a peak loss of 0.610 K_1 occurring at $H = 0.5\,K_1/\mu_0 M_s$ falling to zero at $2K_1/\mu_0 M_s$ (magnetization is confined to a (001) plane). Observed rotational hysteresis losses below the expected threshold fields have been taken as supporting models for incoherent reversal modes. Persistence of losses up to high fields may result from certain symmetries of anisotropy (a unique axis within a unique plane (Owens, 1982)).

Models for rotational hysteresis in multidomain grains are more or less non-existent.

4.3.6 *Particle–particle interactions*

The role of microstructure in controlling the magnetization process is the theme of the present chapter. The importance of some of the elements of microstructure, the size and shape of the grains, the distribution of grain axes and submicroscopic defects have been discussed in the preceding sections. Another important element of microstructure is the geometrical relationship between a magnetic particle and its magnetic neighbours, and their contributions to the magnetic field experienced by the particle. Such particle–particle interactions undoubtedly occur between the particles on magnetic tape (containing 30–40 volume % magnetic material) and probably also between the magnetic mineral grains in rocks and simulated rocks, made by dispersing magnetic powders in a non-magnetic matrix. In the latter two cases the overall concentration of magnetic material may be low ($\sim 1\%$) and the average distance between particles correspondingly high. The distribution of particles may not be uniform, however, either in the natural or synthetic material and local concentrations may be much higher than the average for the sample as a whole. The mean interaction field and its dependence on the overall magnetic state of the assemblage are impossible to model usefully because of the large number of degrees of freedom in the system. It is only in specially constructed systems with few constituents that independent model predictions and experimental observation can be compared. Otherwise the interaction characteristics have to be inferred from magnetic measurements, on the assumption that the interactions exist and are influencing the magnetic properties being measured. This may involve the construction of a Preisach diagram (a two-dimensional histogram showing the coercive force spectrum of an assemblage of particles, the coercive forces in opposite directions for each particle being unequal because of bias due to an interaction field (see, .e.g. Dunlop, 1969)). Another method is to compare the magnetization and demagnetization processes (Cisowski, 1981) the interactions resisting the former and assisting the latter. Alternatively, the difference between observed properties, e.g. hysteresis parameters, and those

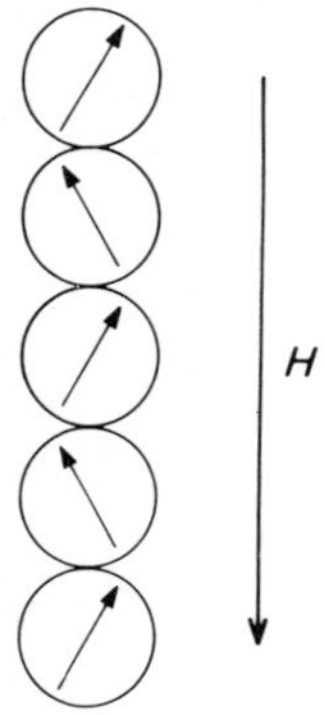

Figure 4.16 The 'chain of spheres' model for magnetization reversal in a system of interacting monodomain particles in which reversal takes place by 'fanning'. The coercive force of the assemblage is lower than that for an isolated monodomain particle reversing by coherent rotation.

predicted by a model, e.g. the Stoner–Wohlfarth model of section 4.3.1, may be ascribed to interactions between particles. One such example is the development of the 'fanning' mode of reversal which may apply in an elongated particle which is not a single unit, in its idealized form, the 'chain-of-spheres' model (Fig. 4.16). In this mechanism the rotation of magnetization within each particle is coherent but the magnetization of the chain as a whole rotates incoherently, the moments of adjacent particles fanning—rotating in opposite senses. It can readily be seen, at least qualitatively, that the dipolar field of neighbouring particles will asist the applied field in achieving reversal of magnetization of the assemblage, at a lower applied field than that necessary to reverse the moment of an isolated particle. This mechanism may be applicable to skeletal or dendritic mineral grains occurring in rocks.

The importance of magnetostatic interactions between particles is likely to be greater in assemblages of particles which themselves derive their anisotropy from magnetostatic effects, i.e. shape anisotropy. Particles which have high anisotropy due, for example, to strain may be less susceptible to the effects of interaction fields. For monodomain particles with shape anisotropy, the relationship between coercive force and volume fraction of dispersed grains, p, $H_c(p) = H_c(0)(1 - p)$ is predicted by several theories and has experimental support.

Interactions between particles of different composition, or different magnetic characteristics such as domain state or magnetic hardness, may produce complex and interesting effects. Interaction between phases with different Curie point temperatures can produce reversed thermoremanence (section 5.4 and 7.3.3). In particles in which variable microstructure or some other mechanism (e.g. subsolvus exsolution) produces regions of hard magnetization surrounded by, or connected with, other regions of soft magnetization, the soft regions will tend to become magnetized by the field due to the magnetic moments of the hard regions. This is analogous to a permanent magnet magnetizing its 'keeper' and the moments of the hard region are 'screened' just as the moment of the magnet is screened. A model expression for the observable moment of a spherical inclusion within a sphere, the two regions having different permeabilities, has been derived by Stephenson (1975) who also considers the variation of the observed moment with temperature when the two regions have different Curie points.

4.4 The time-dependence of magnetization—equilibrium and the effect of thermal agitation

4.4.1 *Equilibrium and disequilibrium*

A demagnetized assemblage of monodomain particles in the presence of a magnetic field, or a magnetized assemblage in zero field are not equilibrium states. Consider an aligned array of uniaxial particles which has been exposed

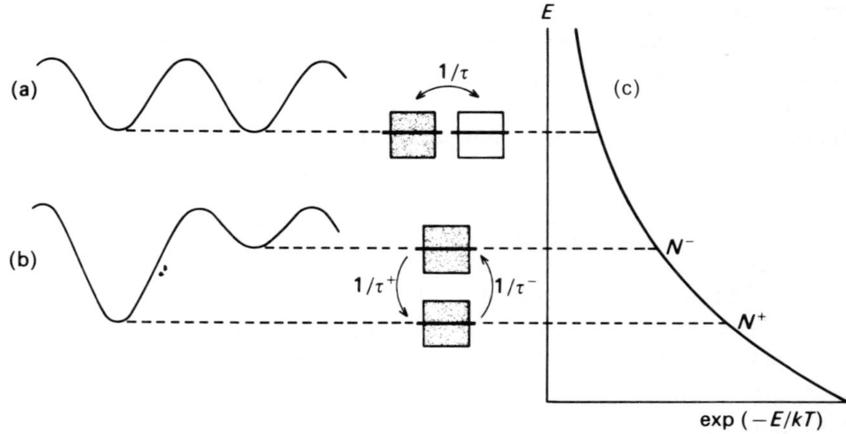

Figure 4.17 (*a*) The distribution of magnetization directions in an aligned assemblage of uniaxial monodomain grains immediately after removal of a saturating field. The two states of equal energy are not equally populated. (*b*) The distribution of magnetization directions of a demagnetized aligned assemblage immediately after application of a field (at $\phi = 0°$) not big enough to remove energy barriers to magnetization reversal. The two states of unequal energies are equally populated. Neither (*a*) nor (*b*) conform to the equilibrium distribution of directions in accordance with the Maxwell–Boltzmann statistics (*c*)

to a saturating field, i.e. large enough to remove anisotropy energy barriers to magnetization change, and the field removed. The state of the assemblage after this irreversible magnetization change may then be represented by Fig. 4.17(*a*) in which the shaded and unshaded boxes indicate the number of particles in the relevant anisotropy energy minimum. Of the two states of equal energy, one is unoccupied. Consider now the same assemblage of grains now demagnetized (Fig. 4.17(*b*)), so that the two orientations are equally occupied, but in the presence of a field smaller than that needed to leave only one minimum in the total energy curve. In this case two states of unequal energy are equally populated. Neither of these two cases corresponds to the equilibrium states of the system in the presence or absence of a field. A similar statement can be made about the distribution of domain volumes and directions in the analogous states of a multidomain grain.

In a system of non-interacting particles in equilibrium, the probability of a particle having energy E is proportional to the Boltzmann factor, $\exp(-E/kT)$, i.e. the population as a whole is distributed among the available energy states according to this distribution function (Fig. 4.17(*c*)). It can be seen that the distribution of particles among the energy levels of neither Fig. 4.17(*a*) nor Fig. 4.17(*b*) conform to the equilibrium distribution. In the first case the two, degenerate, states should be equally populated. In the second, the lower energy state corresponding the minimum nearest to the field direction should have the higher population, the distribution between the two states depending on the temperature (for a given field). Insofar as temperature is a statistical parameter describing the equilibrium distribution of this magnetic system, the temperature of the system after removal of the saturating magnetic field or an application of the non-saturating field becomes undefined. The system is a magnetic thermometer which has not yet come into equilibrium with its environment.

4.4.2 *Approach to equilibrium*

A system in a perturbed, non-equilibrium, state naturally wishes to return to equilibrium or approach a new equilibrium and that is also the case for this magnetic system. Taking first the system left in the magnetized state, after application and removal of a field high enough to bring about irreversible magnetization changes, we see that (by definition) work must be done to reverse these irreversible changes and restore the assemblage to the demagnetized zero field equilibrium condition. The source of the work is the random thermal vibrations, the mechanical 'noise', within each grain. Each grain experiences random thermal agitations with a wide range of amplitudes, the energy spectrum of the agitations depending on temperature. The distribution of thermal vibrations across the energy spectrum is again given by the Boltzmann factor. The probability of an individual grain experiencing a high energy thermal 'kick' is lower than that of receiving a low energy kick and, therefore, on average, each grain would have to wait longer before the high energy thermal kick arrived. The spectrum of thermal vibrations depends on temperature via the Boltzmann factor, the lower the temperature the lower the incidence of the higher energy vibrations. The probability of a thermal vibration of energy amplitude E occurring in a given time interval is proportional to $\exp(-E/kT)$ ($\times$ the time interval) and thus the rate of incidence of such agitations is also proportional to $\exp(-E/kT)$. This rate has the dimensions of $(\text{time})^{-1}$. Thus $1/\tau = f_0 \exp(-E/kT)$ where τ is a characteristic time, the relaxation time—a time in which there is a significant probability of an individual grain receiving a thermal vibration of size E. τ does not describe the experience of an individual grain but rather is a statistic describing the assemblage as a whole. f_0 is a fundamental upper limit to the rate at which the grains can respond to the incoming thermal vibrations. In the present context we are concerned with a magnetic response, the reversal of the magnetization vector of a grain, for which the rate f_0 is taken as $10^9\,(\text{sec})^{-1}$. The random thermal vibrations, of large enough size, will initiate random magnetization reversal and thus allow equilibrium to be approached. Both the equilibrium state and the rate of approach to equilibrium are temperature dependent. Clearly the palaeomagnetic record as carried by minerals in rocks is a non-equilibrium state, so the rate of approach to equilibrium can apparently be slow enough (fortunately for geophysics) that even geological time is not adequate for equilibrium to be reached.

Returning to Fig. 4.17(*a*), the absolute size of the energy barrier opposing magnetization reversal for grains with uniaxial anisotropy constant (from shape, induced or crystal anisotropy sources) K_u and volume v is $K_u v$. The rate at which thermally initiated magnetization reversal will occur is then the product of $f_0 \exp(-K_u v/kT)$ and the number of candidates for reversal, initially all located in the left-hand minimum. The rates at which reversals take place from left to right and right to left do not become equal until the dynamic equilibrium of equal populations of the two levels is reached. The averaged behaviour of the magnetization of the system as a whole follows the 'relaxation' equation $M = M_s \exp(-t/\tau)$. A similar description applies to the assemblage of Fig. 4.17(*b*) but now the barriers for reversal in the two directions are not the same, the left-hand minimum being raised by an amount $\mu_0 M_s H v$, the right-hand minimum being lowered by the same amount and the maximum raised by the field. The reversal from the left to the right in the figure now takes place more rapidly than that from right to left, because of the higher incidence in the former case of thermal kicks of large enough size, i.e. $1/\tau^+ > 1/\tau^-$ in the figure. Eventually the dynamic equilibrium of equal rates of reversal, but unequal populations, is reached according to a relaxation equation of the form $M = M_{eq}(1 - \exp(-t/\tau))$. The relaxation time is now a function of the applied field. M_{eq}, the equilibrium magnetization of the assemblage, depends on the field and temperature and is readily obtained from the equilibrium population of the two levels, $+\mu_0 M_s H v$ and $-\mu_0 M_s H v$, which according to the Maxwell–Boltzmann statistics are proportional to $\exp(-\mu_0 M_s H v/kT)$ and $\exp(\mu_0 M_s H v/kT)$. Designating these numbers as N^- and N^+ respectively, the fractional excess of moments aligned in the field direction is $(N^+ - N^-)/(N^+ + N^-)$ or $\tanh(\mu_0 H_s H v/kT)$. M_{eq} then becomes the magnetization of the material weighted by the statistical function, i.e. $M_s \tanh(\mu_0 M_s H v/kT)$ for this two-state system—an aligned array of uniaxial monodomain grains in a non-saturating field. We see therefore that magnetization changes can take place not only through the process of destruction of anisotropy energy barriers by an applied field but also due to the thermally activated approach to equilibrium of a disequilibriated system. Whether the last process is significant or observable depends on the comparison of the relaxation time and the time scale of interest.

Again the same description and remarks also apply in principle to barriers to domain wall translation in multidomain grains.

4.4.3 *The time scale*

The relaxation time is a strong function of T, v (and K_u) and in practice it often seems possible to divide magnetic behaviour into two separate regimes, one in which relaxation times are shorter than the time scale of interest and in which only equilibrium states are possible; and one in which relaxation times are longer than the relevant time scale and non-equilibrium states are stable. In the latter regime, magnetization takes place only by modification of energy barrier functions—displacement or destruction of minima. In the former the energy barriers only define the states and statistics of the system, hardly influence the magnetization process, and no stable remanence is possible. At the boundary between the two regimes when time scales and relaxation times are similar, both processes contribute to the magnetization process, i.e. displacement of minima by the field followed by thermal excitation over barriers. An observed time dependence of magnetization (or demagnetization) either in mono- or multidomain grains is called 'magnetic viscosity', of which more later. The location of the boundary between the two regimes depends on the selection of a suitable time scale. In laboratory experiments this might be of the order of one minute—on the geological time scale, anything up to 10^9 years. Rearranging the expression for relaxation time to make volume the subject, we obtain $v = (kT/K_u)\ln(f_0\tau)$ which, T and K_u being constant, yields the ratios for $v(\tau)$ where $\tau = 10^2$ s and 10^{16} s, $v(10^2):v(10^{16}) = 1:2.3$. Or, taking the cube root to express this in terms of a ratio of particle dimension, $L(10^2):L(10^{16}) = 1:1.3$. Thus the boundary between the two regimes corresponds to a variation in particle dimension of only 30% even for a wide-ranging choice of time-scales. Thus for any (probably narrower) range of times of interest the boundary volume or dimension is taken as having a specific value, determined only by K_u and T known as the 'blocking volume' v_B. The value of v_B itself will depend on temperature both explicitly and via the temperature dependence of K_u. The boundary between the two regimes in the temperature field can similarly be seen to have the same relative insensitivity to choice of relaxation time, and a 'blocking' or more strictly an 'unblocking temperature' T_B (the present calculation applying to zero field) can be conceived as though the separation between the two regimes was a critical point phenomenon. Taking $\tau = 10^2$ s, the unblocking condition is $K_u v_B = 25\,kT_B$. Below T_B remanent magnetization is stable and is removed only by application of reverse fields and, above T_B, remanence decays away rapidly to zero. The determination or calculation of T_B is more problematic than that of v_B. K_u itself is a function of T but not (or rarely) of v, so that the solution of the relaxation time expression for T is less straightforward, K_u having to be known at all temperature (up to the limit of the Curie point). The division of the regimes in the temperature field, with v constant, lends itself directly to the description of a process in nature, the cooling of magnetic grains in an igneous rock, and leads to the mechanism of acquisition of thermoremanent magnetization (TRM). Division in the particle volume field, with constant temperature, applies directly to the growth of mineral grains a constant temperature in sediments or other rock types and provides the mechanism for chemical remanence (CRM), both mechanisms to be discussed in later chapters.

4.4.4 *Magnetic properties*

So far we have considered only a simple system, an aligned assemblage in low enough fields that the two-state nature of the system is retained. In high fields the minima in the anisotropy energy are overwhelmed and only one minimum remains, in the direction of the field. The absolute amplitude of the magnetostatic energy term is $\mu_0 M_s H v$, now much greater than $K_u v$. It is now magnetostatic and not anisotropy energy which is to be compared to thermal energy. Although Boltzmann statistics still apply, the energy scheme of the system is changed. The magnetization vector of the grain may now make excursions about the energy minimum (as in Fig. 3.7), the necessary work being provided by thermal agitation. Of course if $kT \ll \mu_0 M_s H v$ because, for example, of large v or small T, the magnetization remains in the field direction. Neglecting any residual contribution from the anisotropy term, the energy for the magnetization vector of the grain at angle θ to the field is $-\mu_0 M_s H v \cos\theta$ and the probability of this orientation being occupied, or the average number of grain moments to have this orientation, is proportional to $\exp(\mu_0 M_s H v \cos\theta/kT)$. The resultant moment of the assemblage is derived in the same way as outlined for paramagnetism in section 3.6.1. In the present context it is the moment of the grain as a whole which is being thermally agitated. This is equivalent to very large S or essentially a continuum of allowed orientations, and the Langevin function $\mathscr{L}(\alpha)$ results, where $\mathscr{L}(\alpha) = \coth\alpha - 1/\alpha$. The magnetization of the assemblage is then the material magnetization weighted by the Langevin function with $\alpha \equiv \mu_0 M_s H v/kT$, and has a similar form to the Brillouin function of Fig. 3.13. The Langevin function is tabulated in Appendix 3, p. 176, of Stacey and Banerjee (1974). 90% of the aligned saturation magnetization moment is achieved when $\mu_0 M_s H v/kT = 10$. In small fields $\mathscr{L}(\alpha)$ approximates to $\alpha/3$ and the magnetization per unit volume of magnetic material is $M = \mu_0 M_s^2 v H/3kT$ or for dispersed grains (N per unit volume) the moment per unit volume of the dispersion is $N\mu_0 M_s^2 v^2 H/3kT$. The similarity of this description of the assemblage of small grains to the paramagnetism of an assemblage of dispersed magnetic ions in a dilute material led to the term 'superparamagnetism'. Like paramagnetism, initial susceptibility follows a Curie law (but has a very high value, see below) and high field magnetization curves at different temperatures plot as one curve when magnetization is plotted against (H/T) (like Fig. 3.13). The term superparamagnetism is used to describe the behaviour of any system above its blocking temperature or below its blocking volume. In the high field (or low anisotropy) case the behaviour is the same whether the particles have aligned easy axes or a random alignment. A non-saturating field and a random array of axes presents a complicated problem, the shift of the energy minima depending on field strength and orientation (as for example in Fig. 4.8). However, in low fields the initial susceptibility is the same as in the aligned case.

It is of interest at this point to compare the susceptibilities of a random assemblage of 'stable' (large v, long τ) and superparamagnetic (small v, short τ) monodomain grains. At a fixed temperature the susceptibility of the superparamagnetic assemblage rises linearly with increasing volume. The maximum susceptibility for this state is reached when $v = v_B$ above which the assemblage enters the stable monodomain state with susceptibility $\mu_0 M_s^2/3K_u$ (section 4.3.1). Substituting for v_B, the maximum superparamagnetic susceptibility is $\mu_0 M_s^2 \ln(f_0\tau)/3K_u$ and the ratio between the two simply $\ln(f_0\tau)$. With $\tau = 10^2$ s the assemblage just below the blocking volume has a susceptibility about 25 times higher than that in the stable monodomain state above. The susceptibility is also enhanced on passing through the transition by raising T, v remaining constant. This may contribute to the 'Hopkinson effect'.

Above the blocking volume the monodomain assemblage has the hysteresis properties described in section 4.3.1. Below the blocking volume M_r and H_c are effectively zero. At the lower end of the stable monodomain range, thermal agitations can combine with a reverse field to produce reversal at a lower field than that necessary in the absence of thermal fluctuations (e.g. $2K_u/\mu_0 M_s$ for $\phi = 0/180°$, well above the blocking volume for that temperature). The $\phi = 0/180°$ case and an aligned assemblage are illustrated in Fig. 4.9, curves (*a*) to (*e*). The barrier to reversal is progressively lowered by the reverse field, eventually coming into the range of thermal energy so that an equilibrium distribution with most of the assemblage reversed can be reached in a short time at that temperature (barrier height $\sim 25\,kT$). In effect, the field has reduced the blocking temperature to the temperature at which the magnetization reversal is being observed. The absolute height of the barrier in terms of the reverse field H is the difference between the elevated minimum and the maximum of the energy curve of Fig. 4.9(*c*). This barrier in fact equals $K_u v(1 - \mu_0 M_s H/2K_u)^2$. Equating this to $25kT$ and putting $H = H_c$, the fall in coercive force with increasing temperature or decreasing volume at the stable monodomain/superparamagnetic boundary is given by $H_c = (2K_u/\mu_0 M_s)(1 - 5(kT/K_u v)^{\frac{1}{2}})$ (see also Fig. 6.1). It can be seen that $H_c = 0$ for $K_u v = 25kT$, the condition for superparamagnetic behaviour. M_{rs}/M_s would equal the stable monodomain value, 0.5, for almost the whole transitional range.

4.4.5 *Viscous magnetization*

Towards the lower end of this transitional region, where the volume, and therefore barrier height, are too large, or the temperature is too low for superparamagnetism proper, the approach to equilibrium may be observed without too much difficulty. If $\tau \sim 10^7$ s (the order of several months or one or two years, probably about the upper limit for most, but by no means all, dedicated experimentalists), corresponding to $K_u v = 40kT$ and particle dimension about 10% bigger than the superparamagnetic limit, H_c is still significant at about 20% of the stable monodomain value for $\tau \sim 10^2$ s and $M_{rs}/M_s = 0.5$, when observed for times less than 10^7 s. The material will, however, exhibit magnetic viscosity, i.e. the growth of magnetization with time in small fields to acquire a viscous remanent magnetization (VRM), or the decay of a remanence acquired by slow growth or by other means. In the simplest model assemblage of grains, with identical barriers to magnetization reversal, the growth and decay of magnetization would follow the relaxation equations of section 4.4.2. In more complicated models or in real assemblages, a spectrum of particle barrier heights or volumes and a spectrum of relaxation times must be supposed. The growth and decay curves depend on the form of the spectrum, this form being dictated by nature or chosen by the theoretical modeller. A tractable model spectrum of relaxation times has $\log \tau$ constant between upper and lower limits τ_1 and τ_2 and results in symmetrical viscous acquisition and decay curves, parts of which have the form $M = S \log t + C$. S, the viscosity coefficient for acquisition or decay, depends on the applied field in the acquisition case, the nature of the starting remanence in the decay case and τ_1 and τ_2 in both cases.

Our description of the role of thermal energy in providing the means by which energy barriers resisting approaches to equilibrium are overcome applies also to multidomain grains. In this case the absolute height of the barriers is $\Delta E_w A$ for a wall of area A. At high temperature when kT is large or when the barrier height is low because of a high degree of crystalline perfection in the material, the domain wall may make Barkhausen jumps without the necessity of the barriers being overcome by an applied field. The material is then at or above its unblocking temperature. The presence of a field will bias the jumps by reducing the barriers to motion in one direction and raising them to jumps in the other (see Fig. 5.2). Susceptibility will be enhanced (the Hopkinson effect). The magnetization change accompanying each jump also depends on the submicroscopic microstructure and may be greater in crystals with low concentrations of defects. Again with a suitable combination of parameters, the viscous growth or decay of magnetization at a given temperature will be conveniently observed in laboratory time. In the case of growth of magnetization, the barriers to growth are at first the lower but gradually become higher as the translation of the domain wall produces a net magnetization and the consequent self-energy and demagnetizing field. In the case of a magnetized grain in zero field, the barriers to decay are the lower. The Stacey model (see Stacey and Banerjee, 1974) predicts that magnetization will grow as the logarithm of time, the viscous growth coefficient in this case being proportional to temperature.

4.5 Summary

The resistance to the magnetization process is governed by a combination of the microstructure, principally the particle volume, shape, internal and surface defects, and the properties of the material, principally the anisotropy, intrinsic or extrinsic. The spontaneous magnetization provides the link with the magnetizing field and assists the magnetization process. At a given temperature the behaviour or the state of a grain may be classified as being either monodomain or multidomain and, having decided on a time scale, the monodomain regime may be further divided into superparamagnetic and stable monodomain. The transitional region between the superparamagnetic and stable monodomain regions is associated with a time-dependent viscous magnetization and the boundary between monodomain and multidomain regimes is the so-called pseudo-single-domain range. Figure 4.18 shows the model grain-size variation of magnetic properties at a given temperature, say midway between near-absolute zero temperatures, where thermal energy is negligibly low except for vanishingly small grains, and near-Curie point temperatures, where anisotropy and related parameters become small compared to thermal energy. The low field behaviour is represented by the susceptibility. This rises linearly with volume in the superparamagnetic region (although in very small grains there is the possibility that M_s itself is volume-dependent because of the relative high volume of weakly exchange-coupled atoms near the surface—this would produce a deviation from the linear model). At the superparamagnetic/stable monodomain boundary, the applied field is no longer assisted by thermal agitation and the susceptibility falls by a factor of about 25, (corresponding to a measuring time constant of 10^2 s). The susceptibility then remains constant until the pseudo-single domain region is reached at which it rises to become again approximately constant in the multidomain range. The high field magnetization is constant

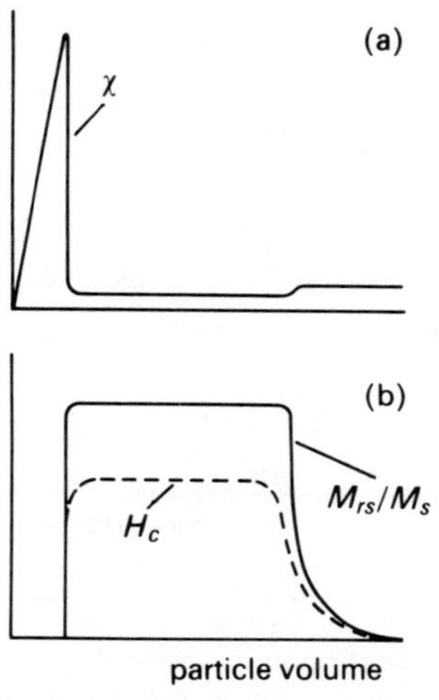

Figure 4.18 (*a*) The variation of susceptibility with particle volume at constant temperature showing the superparamagnetic, stable monodomain and multidomain regimes. (*b*) The variation of M_{rs}/M_s and H_c with particle volume.

throughout the volume range and probably approaches M_s with available laboratory fields for most materials of interest in rock magnetism in the median temperature range. The model hysteresis properties as a function of volume are sketched in Fig. 4.18(*b*). H_c and M_{rs}/M_s are effectively zero in the superparamagnetic region, rise rapidly to the monodomain model values at the onset of stable monodomain behaviour and fall in the incoherent mode/pseudo-single domain region to become approximately constant in the multidomain range. The variation of the magnetic parameters with volume will be different at different temperatures. The superparamagnetic/stable monodomain boundary volume, v_B, varies as T/K_u, but as the variation of K_u with temperature can be quite complicated, depending on the source of the anisotropy, the variation of v_B with T cannot be generalized. For shape anisotropy K_u generally falls as T rises (for a Q-type ferrite at least) and v_B

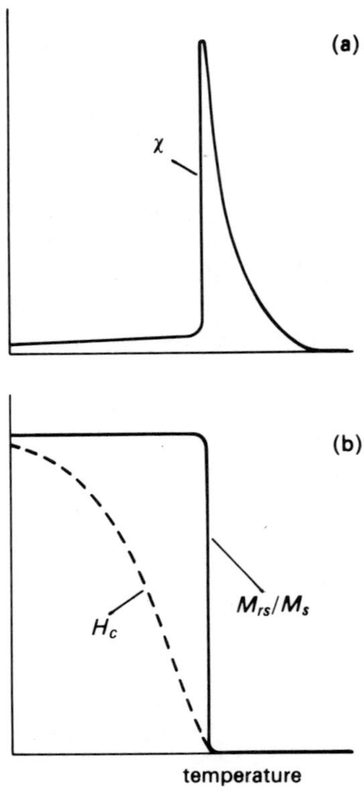

Figure 4.19 (*a*) The variation of susceptibility as a function of temperature for a material in which the anisotropy falls steadily and rapidly as temperature rises and in which the stable monodomain state gives way to superparamagnetism. (*b*) The temperature variation of M_{rs}/M_s and H_c with temperature in the same material with the same particle size. The temperature dependence of anisotropy is quite varied and many possibilities involving multiple transitions between superparamagnetism and the stable monodomain state and between monodomain and multidomain states could be envisaged.

will therefore rise as temperature rises. This result is general for temperatures near the Curie point where anisotropy from whatever source falls as temperature rises. The monodomain/multidomain boundary volume depends on $K^{\frac{1}{2}}/M_s$. If K falls more rapidly with rising temperature than does M_s, the boundary volume will reduce as temperature rises. At high temperature the stable monodomain region may therefore become progressively narrowed.

A plot of the magnetic properties at constant volume, as a function of temperature, can only be made once the variation of material properties is known. Figure 4.19(*a*) shows the susceptibility against temperature for one possible case in which stable monodomain behaviour gives way to superparamagnetism as temperature rises, K_u falling more rapidly than M_s^2 (i.e. not shape anisotropy). The susceptibility rises gradually, jumps at the boundary, T_B, and then falls at a rate faster than the T^{-1} Curie law because M_s is also falling as T rises. Other possibilities are a transition from monodomain to multidomain behaviour or simply a constant multidomain behaviour as temperature rises. In these cases there may be a more or less steady susceptibility rise with temperature, a more rapid rise at the transition in the former case, followed by a rapid fall in the region of the Curie point. The temperature variation of M_{rs}/M_s and H_c for the monodomain to superparamagnetic case is shown in Fig. 4.19(*b*). M_{rs}/M_s remains at the constant model value appropriate to the symmetry of the anisotropy and falls to zero at T_B. H_c falls steadily, at first due to the fall in anisotropy then more rapidly due to the effect of thermal fluctuations, before becoming zero in the superparamagnetic region.

A detailed discussion of the model mechanisms and the magnetization process as actually observed in magnetic minerals must await the detailed description of the material properties of Chapters 7 and 8.

Selected bibliography

*Chikazumi, S. (1964) *Physics of Magnetism*. J. Wiley & Sons Inc., New York, London, Sydney.

Cisowski, S. (1981) Interacting vs. non-interacting single-domain behaviour in natural and synthetic samples. *Phys. Earth Planet. Int.* **26**, 56–62.

Craik, D.J. and McIntyre, D.A. (1967) Critical size of magnetic particles with high uniaxial anisotropy. *Proc. Roy. Soc.* **A302**, 99–112.

Dunlop, D.J. (1969) Preisach diagrams and remanent properties of interacting mondomain grains. *Philos. Mag.* **19**, 369–378.

*Dunlop, D.J. (1973) Theory of magnetic viscosity of lunar and terrestrial rocks. *Rev. Geophys. Space Phys.* **11**, 855–901.

*Dunlop, D.J. (1981) The rock magnetism of fine particles. *Phys. Earth Planet. Int.* **26**, 1–26.

Dunlop, D.J. (1983) On the demagnetizing energy and demagnetizing factor of a multidomain ferromagnetic cube. *Geophys. Res. Lett.* **10**, 79–82.

Eaton, J.A. and Morrish, A.H. (1969) Magnetic domains in hematite at and above the Morin transition. *J. Appl. Phys.* **40**, 3180–3185.

Halgedahl, S. and Fuller, M. (1980) Magnetic domain observations of nucleation processes in fine particles of intermediate titanomagnetite. *Nature* **288**, 70–72.

Halgedahl, S. and Fuller, M. (1981) The dependence of magnetic domain structure upon magnetization state in polycrystalline pyrrhotite. *Phys. Earth Planet. Int.* **26**, 93–97.

*Jacobs, I.S. and Bean, C.P. (1963) 'Fine particles, thin films and exchange anisotropy (effects of finite dimensions and interfaces on the basic properties of ferromagnets),' in *Magnetism* (ed. G.T. Rado and H. Suhl) Academic Press, New York, pp. 271–350.

*Morrish, A.H. (1965) *The Physical Principles of Magnetism.* J. Wiley & Sons Inc., New York, London, Sydney.

Owens, W.H. (1982) A simple model for non-vanishing rotational hysteresis in haematite. *Phys. Earth Planet. Int.* **27**, 106–113.

Rosenberg, M. and Tanasoiu, C. (1975) 'Magnetic domains', in *Magnetic Oxides* (ed. D.J. Craik) J. Wiley & Sons, London, New York, Sydney, Toronto, Chapter 9, pp. 483–573.

*Schmidt, V.A. (1973) A multidomain model of thermoremanence. *Earth Planet. Sci. Lett.* **20**, 440–446.

Soffel, H.C. (1981) Domain structure of natural fine-grained pyrrhotite in a rock matrix (diabase). *Phys. Earth Planet. Int.* **26**, 98–106.

Soffel, H.C. and Appel, E. (1982) Domain structure of a mall synthetic titanomagnetite particles and experiments with IRM and TRM. *Phys. Earth Planet. Int.* **30**, 348–355.

*Stacey, F.D. and Banerjee, S.K. (1974) *The Physical Principles of Rock Magnetism.* Elsevier Scientific Publ. Co., Amsterdam, London, New York.

Stephenson, A. (1975) The observed moment of a magnetized inclusion of high Curie point within a titanomagnetite particle of lower Curie point. *Geophys. J.R. astr. Soc.* **40**, 29–36.

*Source material not specifically cited in context in the text.

5 Thermoremanent magnetization

5.1 The mechanism—definitions of blocking temperature

We now come to the mechanism which provides the physical basis of the palaeomagnetic method as applied to igneous rocks, or other rocks heated without physical or chemical change in nature during metamorphism. Igneous rocks solidify, the magnetic minerals crystallize out and cool. At the Curie points of the magnetic constituents magnetic order sets in and, just below the Curie temperature, the spontaneous magnetization is weak. Anisotropy, from whatever source, is even weaker. Thermal energy is high compared to the barriers to magnetization change and the magnetic configuration in the system quickly reaches the thermal equilibrium corresponding to the presence or absence of an ambient magnetic field. This will be so regardless of the absolute value of the Curie temperature, so that the process would be equally effective for minerals or other materials with Curie points below room temperature, such as ulvöspinel. Although such minerals will not contribute to the natural remanent magnetization of rocks, they may acquire remanence in laboratory experiments aimed at understanding the processes believed to operate in nature, and the properties of the minerals themselves have intrinsic interest. The rapid approach to thermal equilibrium will also take place regardless of the grain size or other microstructural properties. Minerals in the monodomain state will be superparamagnetic regardless of their volume, $K_u v$ being less than kT because K_u is small. Similarly the heights of barriers to domain wall motion depend, via microstructure, on material properties such as anisotropy or magnetostriction and, because these will be small effects near the Curie temperature, magnetization change will take place readily by thermal activation. As the temperature of the igneous rock falls even further, anisotropy, again from whatever source, tends to rise and thermal energy falls. The rate at which equilibrium is approached also falls and may eventually become slow enough to be disregarded. The blocking temperature separates the two regimes; above T_B the equilibrium population of the states will obtain, below T_B it will not. Above T_B, magnetization change takes place by thermal activation over energy barriers; below T_B, magnetization change will take place only in fields large enough to remove anisotropy or wall energy barriers. A system falling through the blocking temperature region stays in equilibrium with the field in the higher region and, on passing into the lower region, is left in the last state in which equilibrium was possible. In the lower region the field becomes irrelevant, and may change in size (within limits) and direction without affecting the distribution of grain magnetization vectors. The field will modify the magnetization only by reversible displacement of energy minima, i.e. via the susceptibility. The magnetization of the system at the lower temperature can only be irreversibly changed by application of fields high enough to destroy the barriers, i.e. of the order of the coercive force. The system has acquired a thermoremanent magnetization, a TRM, in which the magnetization produced by a small field, assisted by thermal energy, requires a much larger field, in the absence of significant thermal energy, for its removal. TRM is essentially confined to fields which are not large enough to bring about irreversible magnetization change isothermally (i.e. by removal of barriers) at the final temperature which the material reaches on its descending path, at which temperature it may have experienced subsequent changes in ambient field direction and strength. A remanence acquired on removal of a field large enough to remove barriers is the isothermal thermoremanence (IRM) appropriate to the temperature of removal. An assemblage which does not have a single-valued energy barrier, such as a random array of identical uniaxial monodomain grains (in the sense that a field between $K_u/\mu_0 M_s$ and $2K_u/\mu_0 M_s$, at the final temperature, would cause irreversible magnetization change in only part of the population of Fig. 4.13), or an array of non-identical grains, or multidomain grains with variation in domain wall barrier heights either among or within grains, may carry both TRM and IRM after removal of the field, provided the field is large enough. The acquisition of the IRM has no connection with the heating process. This is also true of samples acquiring partial TRM (PTRM), to be described later.

The adopted definition of the blocking temperature in the dynamic regime of falling temperature now becomes more complicated. The defining expression of section 4.4.3, the energy barrier $\sim 25kT_B$, followed from the simple adoption of a time-scale of interest of the order of 10^2 s, i.e. a typical time associated with the laboratory study of the magnetization process. It might be emphasized once again that the purpose of introducing the idea of a blocking temperature is to simplify our description of the physical phenomena. Even above the blocking temperature, however defined, equilibrium is reached only in infinite time as can be seen from the relaxation equations of section 4.4.2. The longer the time available, the nearer equilibrium is approached. To reach a given proximity to equilibrium, more time will be needed when energy barriers are high compared to kT, i.e. generally at lower temperatures. In the case of a slowly cooling system more time is available at any given temperature than in a faster cooling system. A given proximity to equilibrium may therefore be reached at a lower temperature.

The lower the temperature, the smaller the distance the system will have travelled towards equilibrium in a given available time. At room temperature the magnetic minerals in a rock are engaged in the journey towards thermal equilibrium with whatever field the rock is experiencing and, given enough time, depending on the energy barriers, will approach equilibrium. During cooling there is no transition temperature above which the system is infinitely flexible and equilibrium is reached, and below which the system becomes irredeemably petrified in the state which obtained on passing through the transition. It may nevertheless be desirable to define a blocking temperature as a simple parameter giving information about the temperature-time history of the system, such history having implications for the interpretation of the remanent magnetization in whatever geophysical or other context. Again a time-scale must be adopted which describes in this case the variation of temperature, and hence energy barriers, with time. Just as in laboratory measurements the time constant of an isothermal measurement provided the criterion, so a cooling time constant may be devised describing the rate of change of temperature with time and the rate of change of energy barriers with temperature. When the relaxation time is of the same order as the cooling time constant, the system is at its blocking temperature. The longer the cooling time constant, the lower the blocking temperature. As the relaxation time is a strong function of temperature both directly and, more important, via its strong dependence on temperature-sensitive energy barriers, a wide range of relaxation time is swept through in a narrow temperature range, and very wide ranges in cooling time constant are needed to significantly affect the blocking temperature. Evaluation of the cooling time constants applicable in slow cooling in nature, or more rapid cooling in laboratory or natural quenching, requires detailed knowledge of the temperature-dependent material parameters.

An alternative approach is to summarize the thermo-magnetic temperature-time history of the system not in terms of a single blocking temperature but in terms of a blocking temperature range, within which equilibrium has been approached to specified degrees. At the upper limit of the blocking range the system closely approaches equilibrium (to within 5%, say), at the lower end the path to equilibrium is begun, but has only reached (say) 5% of the equilibrium magnetization. The blocking range, so defined, again must be determined for a particular material, with its particular temperature dependence of material properties. Figure 9.1 shows the model variation of blocking range with cooling rate for monodomain magnetite with shape anisotropy. For details of the adopted definitions of blocking temperatures or blocking temperature ranges, and their variation with cooling rates, the reader is referred to the papers of York (1978) and Dodson and McClelland-Brown (1980). The blocking temperature range, referred to above, arises from the adopted definition of the blocking temperature, and such a range will be present in an assemblage with identical barriers to magnetization change. The presence of a range of barrier heights in the system will produce, in effect, a range of blocking temperature ranges.

We will now consider the mechanism in more detail and in the models to be discussed, for simplicity, take the blocking temperature to be single-valued for an assemblage of identical grains, i.e. cooling rates are fast or, if not, the adopted T_B is that which produces the same end result as would follow from fast cooling.

5.2 TRM models

5.2.1 *Aligned monodomain grains*

We start with the simplest model, an assemblage of non-interacting monodomain grains having uniaxial anisotropy and aligned easy axes, and with the applied field parallel to the grain axes. Above the blocking temperature, which will depend on absolute barrier heights (and therefore applied field, section 4.4.4) and cooling rate, we will consider equilibrium to be reached. We have already discussed this system (section 4.4.2 and Fig. 4.17) when the field is low enough that two energy minima remain, and have seen that, at equilibrium, the magnetization per unit volume of the magnetic material of the assemblage is $M_s(T)\tanh(\mu_0 M_s(T)Hv/kT)$. The statistical distribution becomes frozen in at $T = T_B$, no further redistribution takes place on cooling to temperature T_0, and the magnetization of the assemblage becomes $M_s(T_0)\tanh(\mu_0 M_s(T_B)Hv/kT_B)$. For low values of the argument, tanh approximates to the argument itself and therefore, for low fields, the remanent magnetization is $\mu_0 M_s(T_0)M_s(T_B)Hv/kT_B$ or $M_s(T_0)aH$ where $a = \mu_0 M_s(T_B)v/kT_B$. We may note that the remanence is proportional to applied field (T_B taken to be independent of field—whence palaeointensity studies) to volume (with, therefore, possible applications in magnetic granulometry) and also records the blocking temperature appropriate to the cooling regime (and applied field). The remanence of this simple highly anisotropic system would, of course, contain little information about the direction of a field not aligned with the grain directions, only the hemisphere in which the field had lain. The result of a decreased cooling rate or an increased field is to lower T_B. $M_s(T_B)$ will generally be increased so the intensity of TRM will be higher for slow cooling rates. The intensity of TRM may be very variable due to the interplay of the factors (v, T_B, $M_s(T_B)$) on which it depends. For the systems and fields of interest in the context of rock magnetism, a TRM would be typically of the order of a few percent of the saturation value, i.e. the assemblage might be regarded as being more or less demagnetized with only a small statistical bias produced by the inducing field.

The model may be elaborated by the inclusion of grains of the same material properties but with different energy barriers by having a variation in grain volume. Each grain size fraction then has its own T_B and the resultant weak-field TRM, for those grains which have $T_B > T_0$, is $M_s(T_0)(\sum w_i a_i)H$ where w_i is the fraction of the blocked assemblage with particle volume v_i.

5.2.2 *A random array of monodomain grains*

The case of a random array of uniaxial particles is more complicated because a field applied at angle ϕ to the grain axis both displaces the minima in the energy curve and changes the energy barrier height. An analytical solution for the equilibrium magnetization as a function of ϕ cannot be obtained and numerical methods would have to be used. If the field is small so that the minima are hardly displaced, i.e. the magnetization of the individual grains remains along their easy axes, the barrier to magnetization reversal can be taken as being the same as if the field were applied along the easy axis, is in the previous section, the field experienced by the grains in each conical annular element now being $H \cos \phi$. The statistical equilibrium for each element corresponds to a net fractional excess of the grains $\tanh(\mu_0 M_s(T)H \cos \phi v/kT)$ aligned in the direction of the positive component of the field. The fraction in each elementary annular cone and its contribution to the resultant magnetization along the field direction is then summed. The necessary integration must be done numerically (Stacey and Banerjee, 1974). Having cooled through the blocking temperature, the system carries a record of the direction of the applied field. The intensity of TRM in

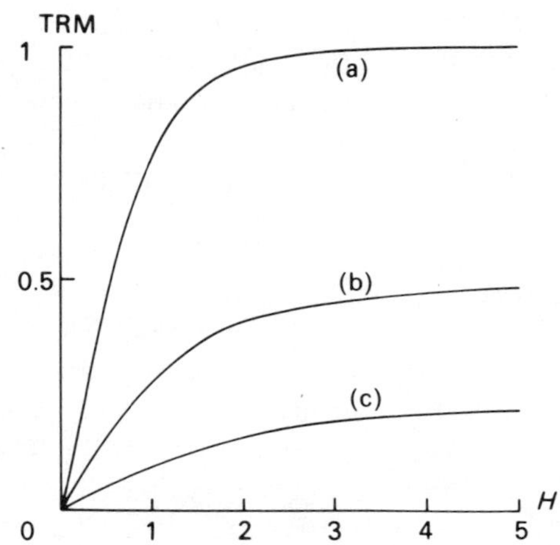

Figure 5.1 Model TRM acquisition curve for uniaxial monodomain grains. (*a*) TRM in units of M_s as a function of field expressed in units of $(1/a)$ (see text) for an array of identical non-interacting aligned grains; (*b*) the acquisition curve for a random array of identical grains; (*c*) the curve for a randomly aligned array of moments uniformly distributed in size between 0 and m_{max}. Here the magnetization is in units of $N_{tot} m_{max}$ and the field in units of $(1/a')$ (see text). In all cases, field independent blocking is assumed. Inter-particle interactions would result in a less steeply ascending curve in each case.

the limit of small field is $(1/3)M_s(T_0)aH$. The form of this model TRM-field acquisition curve for higher fields (and field-independent T_B) is shown in Fig. 5.1 together with the model curve for the aligned array. The remanence saturates for the random array when all vectors are in the same hemisphere about the field direction, i.e. at 0.5 M_s. Again for a range of particle volumes a factor $\sum w_i a_i$ would be included, and the resultant high field acquisition curve would be the sum of a family of curves of the type shown in Fig. 5.1(*b*). The case of a uniform magnitude distribution of monodomain-like moments with identical blocking temperatures is shown in Fig. 5.1(*c*) (see section 5.2.4).

In a further elaboration of the model, the effect of particle-particle interactions may be included. The applied field H experienced by each grain will on average be reduced by the interaction field, this field in turn being a function of the packing factor of the particles and the degree of magnetic alignment in the assemblage at the blocking temperature. The reduction in effective field makes the acquisition curve rise more slowly than that of an equivalent non-interacting assemblage, a higher field being required for saturation.

5.2.3 *Multidomain grains*

In multidomain grains thermal equilibrium is possible when anisotropy related parameters, and hence the barriers to magnetization change, are low. Again the real-world complication of the multidomain state is described in terms of an idealized two-domain rectangular grain with uniform and regular domain wall barriers, and the applied field along the direction of domain magnetization. The final state of the system depends on the interplay of the three, temperature-dependent, energy terms during cooling. The variation of the three energy terms with displacement of the domain wall from the centre of the grain is shown in Fig. 5.2. The magnetostatic energy of the grain (Fig. 5.2(*a*)) due to the external field ($-\mu_0\mathbf{M}.\mathbf{H}v$) varies linearly with domain wall displacement x (see section 4.3.2), the self-energy ($\frac{1}{2}\mu_0 NM^2 v$, N is the effective demagnetizing factor) is parabolic with wall displacement (Fig. 5.2(*b*)), and the fluctuating domain wall energy (Fig. 5.2(*c*)) varies sinusoidally ($-\Delta E_w A \cos kx$) as discussed in section 4.3.3.2. The first term is the only one which depends on field. All three are temperature-dependent, the first through its dependence on $M_s(T)$ and the second on $M_s^2(T)$. The third term will generally depend even more strongly on temperature than the first two, because of the rapid rise of anisotropy and magnetostriction in the region below the Curie point. In a cooling regime the three terms assume importance in the reverse order to the strength of their dependence on temperature. At temperatures just below T_c the magnetostatic term dominates, no energy minima exist, nor do domain walls. The energy of the wall as a function of displacement looks like Fig. 5.2(*a*). As temperature falls further, and $M_s(T)$ rises, the growing self-energy term eventually provides a minimum energy

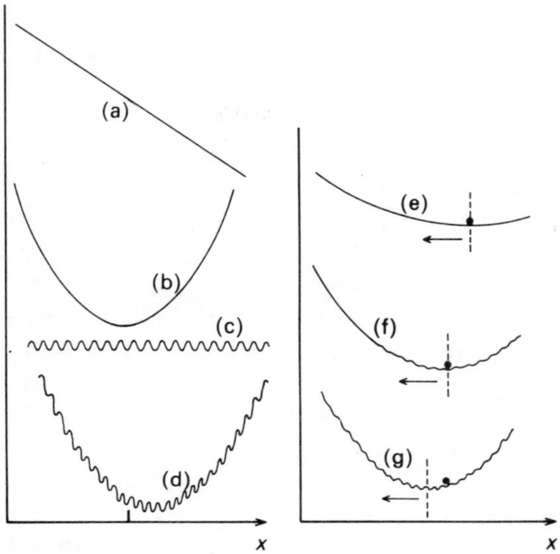

Figure 5.2 The terms contributing to the energy of a domain wall with distance x from the centre of the grain. Curve (a) is the energy due to the applied field; curve (b) the self-energy; curve (c) the fluctuating energy due to variations in microstructure; and curve (d) the total energy. All three terms are temperature-dependent, and curves (e) to (g) (preceded by a curve like (a)) show the growth of the energy terms and the position of the domain wall as temperature falls. (e) corresponds to $T > T_B$, (g) to $T < T_B$ and (f) to, possibly, $T \sim T_B$. (Adapted from Schmidt, 1973).

position within the grain. The displacement of the minimum from the centre of the grain is proportional to field and inversely proportional to $M_s(T)$. As temperature falls and M_s rises still further, the domain wall, having nucleated at the surface, migrates inwards towards the centre (Fig. 5.2(e)). The further fall in temperature leads to the growth and eventual dominance of the fluctuating domain wall energy term. At high temperature, before this stage is reached, the localized minima of the fluctuating term (i.e. at fixed values of x) do not produce corresponding minima on the total energy curve. This is because the amplitude of the fluctuations is so small that its steepest slope ($\Delta E_w Ak$) is less steep than the slope of sum of the other two terms everywhere except near their broad, translating minimum. No localized minima exist, the only contribution of the fluctuating term being a fluctuation in the otherwise constant depth of the mobile minimum. Eventually, the fluctuations become large enough that the localized minima appreciably modulate the broad minimum as it advances towards the centre of the grain (Fig. 5.2(f)). Local minima begin to appear on the leading side and remain approximately fixed in position until they disappear on the trailing side of the broad minimum. At any instant in time there are now several energy minima in which the wall may lie. If thermal agitations are large enough, the wall will spend most time in the deepest minimum, near the centre of the broad minimum. The equilibrium state obtains and the wall will be swept along by the translating broad minimum. As temperature falls, the fluctuating wall energy term approaches the point where it will dominate and control the wall position. Many minima appear superimposed on the broad minimum. The barrier between adjacent minima rises rapidly and thermal energy becomes inadequate for the equilibrium state to be maintained. When $\Delta E_w(T)A$ is the order of $\ln(f_0\tau)kT$, where τ is the appropriate time constant, $T = T_B$ and the domain wall becomes trapped in a local energy minimum. The domain wall is now left behind by the migrating broad minimum and resides in a non-equilibrium position (Fig. 5.2(*g*)). For weak fields, the blocking condition is independent of field, the domain wall is blocked at or near the base of the broader minimum in its position at $T = T_B$. Minimizing the magnetostatic and self-energy terms yields $x_B = LH/2NM_s(T_B)$ (in the notation of sections 4.3.2 and 4.3.3.2) as the blocked position of the domain wall. At the final temperature, T_0, the magnetization per unit volume is $2AM_s(T_0)x_B/AL$, i.e. $M(T_0) = (1/N)(M_s(T_0)/M_s(T_B))H$. Again the intensity is proportional to the field provided T_B, and $M_s(T_B)$, are field-independent. In the monodomain model expression the grain volume appeared both explicitly and implicitly via T_B. In the multidomain expression, the volume or particle dimension only appears implicitly. If the blocking condition is determined only by the interplay of kT and $\Delta E_w(T)A$ then T_B will be a function of particle dimension as, from the model of section 4.3.3.2, $\Delta E_w(T)A \propto F(T)A^{\frac{1}{2}}$ where $F(T)$ is the temperature-dependence of the anisotropy-related properties of the material. A slow cooling rate, resulting in lower T_B and higher $M_s(T_B)$, will produce a reduced TRM intensity.

In higher fields, the blocking condition becomes field-dependent and the TRM intensity is non-linear with field. To determine the form of the field-dependence, a functional dependence of $\Delta E_w(T)$ must be assumed. This may be done indirectly by assuming a form for the $H_c(T)$, $H_c(M_s(T))$, or $\Delta E_w(M_s(T))$ dependences. In the analysis of Schmidt (1973), ΔE_w is taken to vary as $M_s^p(T)$, where $p > 2$, so that the interplay of the three terms can be handled analytically. In high fields the intensity of TRM then varies as $H^{1-[1/(p-1)]}$ until the saturation remanence is produced in a field of the order of the coercive force. The TRM intensity for a given field can also be expressed in terms of the coercive force, yielding $M(T_0) \propto H_c^{1/(p-1)}$. The high field TRM will also have a particle size dependence due to the size-dependence of H_c. A log-log plot of the TRM-field acquisition curve will have three sections (Fig. 5.3), the lowest corresponding to field-independent blocking, the middle section to field-dependent blocking and the uppermost to saturation.

The removal of the field and the magnetostatic term at $T = T_0$ allows a final motion of the domain wall (apart from that corresponding to an isothermal component of remanence acquired because there is a variety of barrier heights and some barriers were destroyed by the field). This final

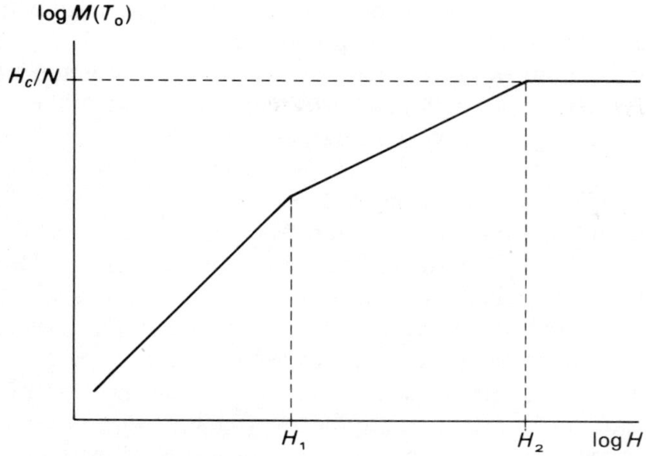

Figure 5.3 Model TRM acquisition curve for a two-domain grain. In the low field region of field-independent thermal blocking, the TRM is proportional to field. Between H_1 and H_2 the blocking is controlled by the applied field. When fluctuating domain wall energy is proportional to $M_s^p(T)$, the slope of this section, plotted bilogarithmically, is $1 - [1/(p-1)]$ (Schmidt, 1973).

adjustment in position of the wall can be described in terms of the reversible parallel susceptibility at T_0 (section 4.3.2). The question of a relaxation of a reversible magnetization did not occur in the monodomain models as the susceptibility is zero for the aligned case and the random array model suffered the restriction that the grain moments were not allowed to deviate from the grain axes. A random array of multidomain grains will record the direction of the inducing field with a specific intensity of remanence 1/3 of that for a grain with domain directions aligned with the field.

5.2.4 *TRM near the monodomain/multidomain boundary*

Taking first the monodomain end of this boundary region where a transitory or semi-stable incoherence (section 4.3.3.1) may facilitate magnetization reversal, in the weak fields of interest in the context of TRM the reversal with transitory incoherence will probably have negligible influence on the blocking condition and the resultant TRM. The temperature-dependence of E_w may favour the development of either a semi-stable incoherence in grains with moments opposed to an applied field direction, or even allow a stable multi-domain state at higher temperature. In the former case, whether TRM differs from that acquired by the equivalent monodomain assemblage depends on whether the partially demagnetized state of the reversely magnetized grains persists as temperature falls below the blocking temperature. A point supporting this would be that the grains are at the upper end of the monodomain range with large volume and high blocking temperature. On the other hand, the effect, if present in the weak fields of interest, may only be possible in a narrow range below the Curie point where E_w is small. If the frozen equilibrium does correspond to the distribution between coherently magnetized grains in the field direction and incoherently magnetized grains opposed to the field, the net magnetization at room temperature, when the semi-stable incoherence has relaxed leaving a monodomain assemblage, will be reduced. If the spin structure reverts to monodomain above the blocking temperature the TRM should be as per the monodomain model.

If the assemblage undergoes a monodomain/multidomain transition at high temperature (but below the blocking temperature), the TRM resides initially in multidomain grains with a non-centred domain wall (in our two-domain model) and net moment in the field direction. On cooling through the transition the domain wall wishes to reduce its area to zero. Presumably in the rectangular grain it takes the shortest route, at the same time leaving the resultant monodomain grain in the lower energy state in the presence of the field. It appears that the assemblage would emerge with an enhanced, possibly saturated TRM. Modelling of this mechanism with the temperature-dependent material parameters of the relevant mineral systems may be worthwhile.

Coming now to multidomain grains in which the multidomain state is stable in zero field and throughout the temperature range of interest, the question to be asked is whether the difficulties that small grains experience in nucleating a wall, the wall having been removed by a saturating field (as in Fig. 5.2(*a*)), will persist to high temperature. The weak applied field resists nucleation and the demagnetizing field assists (Fig. 5.2(*e*)). A minimum occurs in the magnetostatic and self-energies as temperature falls when $M_s(T) > H/2N$. The smaller the field, the higher the temperature at which the wall wishes to nucleate. The higher the temperature, the lower the barrier to nucleation and it would seem that except for grains with extremely small demagnetizing factor in the domain direction (Brown's nucleation condition—see Halgedahl and Fuller, 1980) being $N \geqslant 2K(T)/\mu_0 M_s^2(T)$) nucleation should take place readily even in the absence of helpful surface imperfections. It seems, therefore, that the two-domain model is applicable down to the monodomain/multidomain boundary.

Small grains on the multidomain side of the boundary may be associated with irremovable moments (pseudo-single domain effects, section 4.3.3.3). These moments, of whatever origin, may be treated as monodomain two-state systems, the moments being able to lie parallel or antiparallel to an applied field, or its component along the preferred axis of the moment. The moments are no longer $M_s(T)v$, i.e. that of a uniformly magnetized grain of volume v, but arise from Barkhausen discreteness, the moment associated with the spin structure of a domain wall or due to the unequal volumes of opposed domains in small grains with odd numbers of domains. A random array of identical monodomain-like moments will acquire a weak-field TRM of $(1/3)Nm^2M_s(T_B)H/kT_BM_s(T_0)$ where N is the number of moments per unit volume and m the magnitude of the moments at T_0. For an assemblage having a range of magnitude of moments varying from zero to m_{max} where the number of moments in each magnitude interval is constant (Stacey and Banerjee, 1974) and the total number of moments is N_{tot}, the low field TRM becomes $(1/9)N_{tot}m_{max}^2 M_s(T_B)H/kT_BM_s(T_0)$. The high field expression again requires a numerical integration (Appendix 2 of Stacey and Banerjee, 1974) and produces a TRM-H induction curve which does not saturate sharply in low fields when plausible values of parameters are substituted. The acquisition curve is plotted in Fig. 5.1 with an arbitrarily chosen $N_{tot}m_{max}$ and the field expressed in units of $(1/a')$ where $a' = m_{max}M_s(T_B)/kT_BM_s(T_0)$.

Turning now to the origin of these monodomain-like moments, if they originate in the wall spin structure and are a function of the relative areas of the surface of the grain at which the domain wall edge is exposed and the total surface area of the grain, it may be shown that $N_{tot} \propto 1/L$. The expression for the TRM due to such moments will, therefore, carry an explicit grain size dependence. The irremovable two-state, monodomain-like moments of grains with odd numbers of domains are $\pm 0.125\, M_s$ for three domains and smaller for higher odd numbers, giving a step-wise decrease with increasing L.

5.3 Thermal demagnetization—partial TRM

Alternating field (AF) demagnetization was discussed in section 4.3.4. The stability of a TRM to AF demagnetization is simply governed by the coercive force of the TRM-carrying grains at the temperature at which the demagnetization is to be carried out. For stable monodomain grains this depends on material properties. These are in general well-behaved, i.e. anisotropy and magnetostriction constants and spontaneous magnetization increase steadily as temperature falls, although this is not always the case (e.g. the compensation temperatures in magnetization and anisotropy discussed in Chapter 3). The same remark applies to multidomain grains. A low blocking temperature (i.e. significantly lower than the Curie temperature but above room temperature) may imply a reduction in coercive force due to thermal fluctuations at room temperature, but this depends on how rapidly anisotropy rises as temperature falls. An assemblage of grains which only differ in their volumes and therefore have a range of blocking temperatures may exhibit a significantly wide range of resistance to AF demagnetization, those with the highest blocking temperature being the most stable (see also Fig. 6.1). An assemblage which contains differences in material properties, anisotropy, magnetization and Curie temperature may show no, or even an anti-, correlation between blocking temperature and coercive force at room temperature. This is illustrated by (say) a mixture of titanomagnetites with varying Al concentration (Chapter 7) (see also section 9.1.1).

The demagnetized state is the equilibrium state in zero field. In thermal demagnetization the equilibrium state is made accessible by raising the temperature to the unblocking region. In zero field all orientations in a random array of monodomain grains are equally populated and, once freed from localized minima, the domain wall of our two-domain grain gravitates to the centre of the grain due to the self-energy parabola. On cooling back to room temperature in zero field, a stable demagnetized state obtains. The unblocking temperature as determined by thermal demagnetization is not necessarily the same as the temperature at which the removed remanence was blocked in. The unblocking temperature corresponds to zero field and the time constant of the demagnetization process, i.e. the time the experimenter is prepared to wait at elevated temperature. The blocking temperature of the removed remanence depends on the field and the cooling time constant, if a TRM, or otherwise some other time constant appropriate to the acquisition process. Any remanent magnetization, of whatever provenance, has an unblocking temperature. Thermal demagnetization may be carried out progressively by the stepwise measurement of the remanence at room temperature after each zero field heating to progressively higher temperatures. An assemblage with a range of absolute energy barriers to magnetization change will progressively lose its remanence as the unblocking temperature of each fraction is reached. In the absence of interparticle interactions, each fraction is ideally independent of all other fractions and gains or loses remanent magnetization independently. In the continuous method of thermal demagnetization, the remanence is measured at the elevated temperature and, although this may save experimental time, reduces the time available for chemical alteration at high temperature and may avoid the re-acquisition of TRM during the cooling operation in an inadequately cancelled ambient field, suffers the disadvantage that the residual remanence is weakened at the elevated temperature due to the fall in $M_s(T)$.

The total TRM for a field H may be defined as the TRM acquired when the assemblage is cooled from the Curie temperature (i.e. above the highest possible blocking temperature) to room temperature (T_0) in that field. This should exclude the moment of any grains with coercive force less than H at room temperature which are isothermally blocked. Any grains with blocking temperature below room temperature are superparamagnetic and contribute only to the susceptibility of the assemblage and not to the remanence. A partial TRM, PTRM, arises during cooling when the temperature interval through which the field is applied is less than $(T_c - T_0)$, the field being zero for any other part of the temperature range through which the assemblage cools. If the field is applied, to a previously demagnetized assemblage, on cooling through the temperature interval $(T_2 - T_1)$ grains with blocking temperatures in that range will acquire TRM. Those with blocking temperatures above T_2 will be unaffected either because that temperature field was not entered or, if it was, because the field was zero.

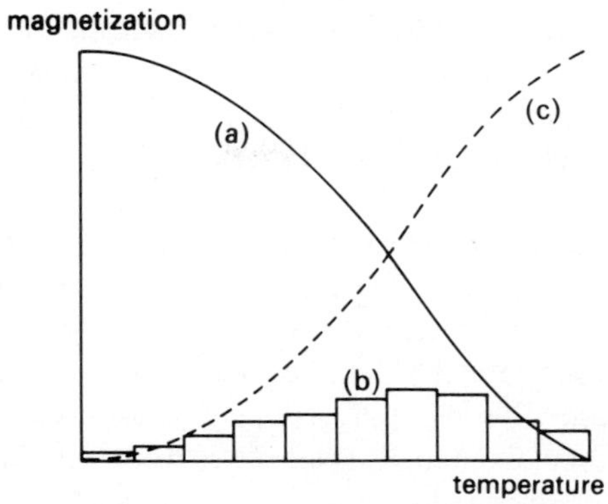

Figure 5.4 Comparison of the progressive thermal demagnetization of total TRM and the PTRMs acquired in the same field. Curve (*a*) shows the residual remanence after heating to temperature *T* and cooling to room temperature in zero field. The histogram (*b*) shows the remanence acquired on cooling through the indicated temperature interval with field applied, the field being zero outside the interval. When all remanences are measured at the same temperature (at room temperature in stepwise experiments) the sum of the PTRMs equals the total TRM. The cumulative PTRM curve (or the PTRM corresponding to a progressively expanded temperature interval) (*c*) is the mirror image of the demagnetization curve.

Any grains with blocking temperatures below T_1 will cool in zero field and stay demagnetized. The independence of each blocking fraction leads to the Law of Additivity of PTRMs, that the sum of the PTRMs acquired independently in a number of adjacent temperature intervals is the same as the single PTRM acquired over the whole interval. This means that the cooling and heating in zero field intermediate between acquisition of each of the set of PTRMs, obtained in any order, has no effect on remanence, as indeed it should not, given the practical validity in the laboratory of the simple well-defined blocking temperature. A thermal demagnetization curve (residual remanence against temperature) of a total TRM is simply related to the (fixed interval) PTRM acquisition histogram or blocking temperature spectrum (Fig. 5.4). One is the derivative of the other, provided the unblocking and blocking temperatures are the same. The cumulative (or expanding interval) PTRM acquisition curve is the mirror image of the demagnetization curve. If PTRMs are imparted to an already magnetized assemblage, the record carried by the fraction in the relevant blocking interval is erased and re-imprinted by the new field in the one operation.

5.4 Self-reversed TRM

Self-reversal of remanence is said to have occurred when the remanence, as measured, is in the opposite direction to the (weak) field which combined with the perturbation characteristic of the remanence impressing mechanism. Many scenarios for self-reversal have been written which could apply in principle to any of the remanence-inducing mechanisms. A quirk of history resulted in the self-reversal phenomenon attracting interest probably out of all proportion to the importance of its contribution to the magnetism of rocks. In the present section we consider only mechanisms which may apply in those magnetic systems which occur in nature and which could produce a reversal of total TRM above room temperature. Some mechanisms require special material properties, others special microstructure.

The N-type ferrimagnet of Fig. 3.16 provides an example of the former category. We suppose the blocking temperature range lies entirely above the compensation temperature at which the macroscopic magnetization vanishes. On cooling through the blocking range, the assemblage acquires a net magnetization in the fixed direction, each grain having its macroscopic moment along a preferred direction within the grain. The ferrite has (say) an *A*-sublattice predominant moment above the compensation temperature and the magnetization of this sublattice is parallel to the macroscopic moment of each grain (Fig. 3.6). On further cooling to the compensation temperature, the magnetization of the two sublattices becomes equal and the macroscopic moment disappears, as does the self-energy. The spin structure is still intact within the grain and still linked to the crystallographic directions by crystal anisotropy. A multidomain grain now gains no advantage from the presence of domain walls which, if they are free to do so, will combine with neighbouring walls of opposite sense or migrate to the surface and disappear. Alternatively the walls may migrate to new equilibrium positions appropriate to an antiferromagnetic substance. On further cooling, the *B*-sublattice moment dominates and the macroscopic moment grows. The domain structure, if lost or transformed, reverts to that needed to reduce the self-energy. It may be that the ambient field, unchanged in direction throughout, influences this process, in which case the resultant moment may be in the same direction as the original TRM. However, if the original domain structure is preserved through the compensation point, the spontaneous magnetization within the domains having reversed, the material now carries remanence reversed in direction.

The monodomain case is more straightforward, provided shape anisotropy is not the factor stabilizing the remanence. If so, the moment of each grain becomes uncoupled from (say) the needle axis near the compensation temperature and will either rotate to the preferred direction consistent with other sources of anisotropy or become superparamagnetic. As macroscopic magnetization grows on cooling further, the superparamagnetic grain will simply re-acquire TRM in the unchanged field. A grain which retained anisotropy will undergo a magnetization rotation back to the grain shape axis. The resultant moment could easily be envisaged to be reversed with respect to the original direction. If the non-shape anisotropy holds the spin structure steadily throughout, a reversed moment results.

A change in composition, due to oxidation, say, could bring about part of this sequence of events isothermally with the same result.

Interactions between phases of different material properties provide a route to self-reversed TRM. In its simplest form a self-reversing system of this kind would consist of grains of two phases in close proximity, having different Curie points. On cooling in an ambient field the grains with higher Curie point (phase 1) and higher blocking temperature acquire individual TRMs if they are multidomain, or the set as a whole acquires TRM if they are monodomain. The TRM is in the ambient field direction and the material of the lower Curie temperature phase (phase 2) has yet to become magnetically ordered. When the temperature reaches the blocking point of phase 2, it also acquires remanence in the field experienced by each grain or the average field experienced by a monodomain set. This field consists of the ambient field plus the interaction field due to the neighbouring phase 1 particles. The interaction field may oppose the ambient field if inter-grain geometry and field direction allow this; most effectively if the phase 2 particles sit broadside-on to the moments of the phase 1 grains. The interaction field will be intense if the inter-particle distance is small and will be a kind of amplified reversed version of the ambient field if, in the temperature interval between the blocking of phases 1 and 2, the spontaneous magnetization of phase 1, and therefore its remanence, have increased markedly. The net field experienced by phase 2 grains may then be reversed as will be the phase 2 TRM. On further cooling to room temperature, the phase 2 TRM can be expected to become enhanced and the observed remanence of the system as a whole may be reversed. This probably requires that both the volume and the spontaneous magnetization of phase 2 exceed those of phase 1 at room temperature. In such a system, the PTRM acquired in a field of any magnitude will be reversed if the field is applied only when phase 1 is acquiring TRM and removed in the lower interval so that phase 2 becomes blocked only in the interaction field. The total TRM in weak field is reversed and remains so for a range of applied field, the remanence at first increasing and then falling as the field increases. With higher applied field, the remanence of the system becomes normal when the applied field exceeds the maximum interaction field. In the reversed state the normal and reverse components in the system are revealed during thermal demagnetization. Examples of self-reversal due to this magnetostatic interaction between phases are given in section 7.3.3.

A second mechanism which relies on a special microstructure and material properties, invokes exchange interactions between magnetic atoms or ions at the adjacent surfaces of different phases. The phases in this case are not so distinct as to be separated by grain boundaries but are rather due to composition gradients with the material. It is supposed that the different regions within the crystal can have different ordering temperatures, the phase with the highest ordering temperature being dispersed in regions of small volume throughout the crystal, resulting in a large interfacial surface between these regions and the bulk of the crystal. This high ordering temperature phase acquires TRM in the field during cooling. If both phases are ferrimagnetic and negative exchange interactions are preponderant, a reversed remanence in the bulk of the crystal requires that, at the interface, atoms in the dominant sublattices of both phases are linked via oxygens. As the lower temperature phase orders, the exchange coupling exerts a powerful influence on the orientations of the sublattices and the moment of the bulk of the crystal which, if monodomain, is reversed. If the bulk of the crystal is in the multi-domain state, the interphase exchange interactions will only influence the domain orientations adjacent to the first-ordered phase. The positions of the walls and the remanence of the bulk phase will be determined by magnetostatics. The case for interphase exchange interactions in non-reversing systems appears to be fairly strong. The main argument for the self-reversing two phase exchange coupled mechanism seems to be the very large field needed to suppress self-reversal in the haemo-ilmenite system (section 8.2.3). However, there appears to be no difference in kind between the behaviour of a suitable devised magnetostatistically interacting monodomain system and a hypothetical exchange coupled system (Stavn and Morrish, 1979) and such a mechanism makes less strong demands on the exact atomic arrangements at the interface.

5.5 Summary

The barriers to magnetization change depend on anisotropy and particle size, and fall as temperature rises towards the Curie point. A weak field, which at ordinary ambient temperatures does not significantly modify the energy barriers and bring about irreversible magnetization changes, may combine with thermal energy at elevated temperature, above the blocking temperature, to magnetize the system by thermal activation over the lowered barriers. The magnetization retained by the system on cooling through the blocking temperature, below which thermal activation is ineffective, is a non-equilibrium state when the field may have changed in size and direction at room temperature. However, the approach to a new equilibrium may be so slow that, depending on the time scale of interest, the magnetization is permanent provided the system is not exposed to fields high enough to overcome barriers, or to reheating above the blocking temperature. The blocking temperature itself requires a conventional definition. The impressed TRM records the direction of the inducing field in an isotropic assemblage and its intensity is proportional to field in low fields.

In a system containing a range of blocking temperatures, each blocking temperature fraction acquires TRM independently provided the fractions are non-interacting. The fraction of the total TRM acquired in a given blocking temperature range is a partial TRM. The sum of the partial TRMs equals the total TRM. The thermal demagnetization characteristics and the acquisition of PTRMs are directly related.

Particle interactions between phases with different blocking temperatures remove the simple law of additivity of PTRMs and may produce a total TRM which is reversed relative to the inducing field. Self-reversal of TRM can also occur in *N*-type ferrimagnets.

Selected bibliography

Dodson, M.H. and McClelland-Brown, E. (1980) Magnetic blocking temperatures during slow cooling. *J. Geophys. Res.* **85**, 2625–2637.

*Dunlop, D.J. (ed.) (1977) Origin of thermoremanent magnetization. *J. Geomagn. Geoelectr.* **29**, (4) (special issue).

Halgedahl, S.L. and Fuller, M. (1980) Magnetic domain observations of nucleation processes in fine particles of intermediate titanomagnetite. *Nature* **288**, 70–72.

Stacey, F.D. and Banerjee, S.K. (1974) *The Physical Principles of Rock Magnetism.* Elsevier, Amsterdam.

Stavn, M.J. and Morrish, A.H. (1979) Magnetization of a two-component Stoner–Wohlfarth particle. *IEEE Trans. Magn.* **MAG 15**, 1235–1240.

Schmidt, V.A. (1973) A multidomain model of thermoremanence. *Earth Planet. Sci. Lett.* **29**, 145–154.

York, D. (1978) Magnetic blocking temperature. *Earth Planet. Sci. Lett.* **39**, 94–97.

*Source material not specifically cited in text

6 Other remanence-inducing mechanisms

6.1 Mechanisms in nature

The TRM mechanism of Chapter 5 may account for the stable palaeomagnetic record carried by igneous rocks or the pre-existing country rock heated during igneous activity. Many sedimentary rock types also carry a stable though weak remanence which provides a useful contribution to the record of the history of the Earth's crust and the ancient field. Mechanisms believed to operate in sediments and to be responsible for their stable remanence are those which lead to the acquisition of chemical remanent magnetization (CRM) and depositional remanent magnetization (DRM). Neither is quite so straightforward or simple as the TRM acquisition process which requires only a change in temperature. A simple reheating may reverse or reproduce the process in the ideal system. A CRM- or DRM-carrying system is unlikely to allow this in practice. CRM is not necessarily confined to sediments. Apart from minerals crystallizing in a solidifying magma above their Curie points and eventually acquiring TRM, an assemblage of grains produced during alteration at temperatures lower than their Curie point may become part-blocked isothermally at the alteration temperature by the CRM mechanism and, conceivably, part-blocked during cooling to ambient temperature producing additional TRM. A sediment which does not fluctuate significantly in temperature is unlikely to acquire TRM. Although some igneous rocks have features in common with sediments, the temperatures at which mineral grains separate out and are deposited within a semi-molten body are invariably above the highest Curie temperatures of materials and DRM is not acquired.

6.1.1 *Chemical remanent magnetization*

The model CRM mechanism is identical to TRM in that thermal activation is effective early in the sequence of events and equilibrium states obtain. At a later stage the system cannot come into equilibrium with field and temperature and is left in a state which is a relic of its earlier history. The magnetic phase is supposed as being produced by precipitation from a liquid or solid solution or from a solid which is reacting with a gas or liquid, the process taking place isothermally at the alteration temperature T_a ($T_a < T_c$; T_a will be near T_0 in a sediment with simple thermal history). The magnetic product nucleates and the grains grow in volume with time. When the grains are small, they are in the monodomain state and the absolute barriers to magnetization reversal, $K_u(T_a)v$, are low compared to thermal energy. In the first and simplest model of aligned uniaxial grains, the statistical function giving the fractional excess of magnetization in the field direction is, as before, $\tanh(\mu_0 M_s(T_a)Hv/kT_a)$. The height of the energy barriers grows with time along with v, as does the relaxation time of the approach to equilibrium. The relaxation time is a strong function of v and, as discussed in section 4.4.3, it is possible to define a blocking volume v_B above which the system approaches equilibrium so slowly that any change in external field direction and intensity has little or no effect on the implanted magnetization. The blocking volume may be defined as that corresponding to a time constant which is equal to the time constant describing the rate of grain growth and which will depend on those factors controlling the reaction kinetics. At the blocking volume, $K_u(T_a)v_B = \ln(f_0\tau)kT_a$, where τ is the reaction time constant and the effect of the field on the blocking condition is ignored. On further growth above v_B, the system enters the regime of non-equilibrium and the statistical distribution of grain moments is frozen into the configuration which the system held on entering the zone. The remanent magnetization of the system per unit volume of magnetic material with grain volume now greater than v_B, is $M_s(T_a)\tanh(\mu_0 M_s(T_a)Hv_B/kT_a)$. If the alteration takes place at some elevated temperature and the system is eventually quenched to ambient temperature T_0, the remanence becomes $M_s(T_0)\tanh(\mu_0 M_s(T_a)Hv_B/kT_a)$ or in weak fields $\mu_0 M_s(T_0)M_s(T_a)Hv_B/kT_a$. Grains with volumes less than v_B but with blocking temperatures between T_0 and T_a acquire TRM during the rise in $K_u(T)v$ as temperature falls. (This does not, of course, apply when $T_a \sim T_0$.) On heating, these grains lose remanence at $T < T_a$ whereas those with CRM will be unblocked at temperatures above T_a and, on cooling in a field, will acquire TRM at these higher temperatures.

It is of interest to compare the intensities of remanence acquired in identical assemblages by the two processes. We may at the same time consider the case of a random array of particles which differs from the aligned case, when the field is weak, only by the inclusion of a factor of 1/3 for both CRM and TRM. The complete CRM induction curve for the random array must be evaluated numerically and is the same as that shown in Fig. 5.1(*b*) where the parameter a is now equal to $\mu_0 M_s(T_a)v_B/kT_a$. The weak field ratio of CRM to TRM is $(M_s(T_a)v_B T_B)/(M_s(T_B)vT_a)$. The CRM is therefore likely to be of lower intensity but will be of identical stability to the TRM, each annular shell of the monodomain array contributing in the same proportion to the total remanence in either case. The CRM/TRM intensity ratio can also be expressed in terms of the material parameters and the reaction and cooling time constants by substituting for v_B and T_B using the respective blocking conditions.

Further growth of the grains may take them out of the monodomain region, i.e. a monodomain/multidomain transition may take place in the presence of a field. The converse case, a multidomain/monodomain transition, was discussed in the previous chapter. In the previous case the remanence became enhanced, here it will become diminished. Initially a domain wall will nucleate at the surface. The self-energy term attempts to drive it into the interior. However, unlike the two-domain TRM model, the local wall energy minima are well established at the lower temperature of alteration where anisotropy-related material parameters have significantly large values. The inward migration is therefore strongly resisted and the CRM of the two-domain grain may then be appreciably larger than a TRM carried by the same grain and may be as high as the saturation remanence value of the grain at that temperature. Further growth leads to the nucleation of more domains and the magnetic moment of the grain must gradually fall.

Just as in the case of TRM, we may define saturation CRM, total CRM and partial CRM. Saturation describes the high field equilibrium state of a single blocking fraction. This is related only to field-dependence and we may therefore have saturation total TRM and saturation partial TRM. 'Total' and 'partial' refer to the blocking or perturbation dependence. In a system with heterogeneous reaction kinetics (quite possible in a gas/solid reaction where the local interface governs reaction rate) the growing grains have a range of volume and at any instant a fraction are passing through the blocking point. A partial CRM is acquired in an interval of time Δt in the field applied during that time. The law of additivity would apply to PCRMs of monodomain grains when, in each time interval, the grains block independently. Whether some process of chemical demagnetization, by which the grains are reduced in volume with time, can be devised so that the simple relation between the PCRM and chemical demagnetization processes could be illustrated, depends on the availability of some suitably simple real-world system. Such a system may be rather remote from those occurring naturally. Total CRM results when long enough time has elapsed that reaction rates become negligibly slow.

Chemical change in the presence of a field is not necessarily accompanied by the acquisition of CRM unless the chemical change is accompanied by a changing height of energy barriers, the barriers at some stage passing through the blocking size. Epitaxial growth on the surfaces of a stably magnetized assemblage of monodomain grains carrying the record of an earlier field would not result in CRM acquisition in the contemporaneous field. The magnetization of the new material might be said to be a CRM controlled by the 'exchange field' of the pre-existing grain. The new and old material are not distinct and separated by a distinct phase boundary. Concentration gradients may be reduced by atomic diffusion and the material properties of the grain may change. The archetypal CRM model system acquires remanence by a change of microstructure with time. A system of the type presently under discussion does not undergo great microstructural changes. The time dependence of material properties can however result in a CRM. Suppose the chemical process brings about a progressive increase in Curie temperature (like the maghemitization process, section 7.2.1) and that initially $T_c < T_a$. As time passes, the aligned assemblage finds itself (at T_a) just below its Curie point, energy barriers $K_u(t)v$ are low and equilibrium obtains. After the barriers have risen, the frozen-in state has magnetization $M_s(t, T_a)\tanh(\mu_0 M_s(t_B, T_a)vH/kT_a)$ (neglecting the change of v with time) with the corresponding weak-field random array expression. t_B is the time at which blocking takes place where $K_u(t_B, T_a)v = \ln(f_0\tau)kT_a$; t is the analogue of the 'fraction of reaction' (z in the case of maghemitization) or the chemical composition of the material at time t. If the initial composition has $T_c > T_a$ the assemblage already has a stable remanence recording a previous field. If $K_u(t, T_a)$ rises with time, the statistical distribution will remain unaltered and the record of the previous field direction will be faithfully retained (on a defined time scale) regardless of the direction of the contemporaneous (weak) field. The intensity of remanence will change, following the change in $M_s(t, T_a)$, the prefix to the statistical function. If $K_u(t, T_a)$ falls with time so that the grains become unblocked, the record of the previous field is lost and the assemblage becomes magnetized, superparamagnetically, in the ambient field but can retain no record of it unless $K_u(t, T_a)$ corresponds to an isotropic point either directly, due to a compensation in competing anisotropies, or indirectly via shape anisotropy and a compensation in magnetization. In this case barriers grow again as time passes and CRM in the ambient field results.

6.1.2 *Depositional remanent magnetization*

The perturbation which facilitates the approach to equilibrium with an ambient field in the DRM mechanism is the buoyancy or lubrication provided by the water in which a sediment is being deposited. Stable monodomain grains or multidomain grains which have previously been magnetized are envisaged as settling out from the suspension in which they have been carried from the source area. The grains experience a torque due to the ambient field and rotate freely into the field direction, the viscous drag resisting rotation being small. Perfect alignment may be disturbed by the random motion of the water molecules giving rise to the Brownian motions of small suspended particles. This is equivalent to thermal agitations of isotropic grains and the degree of alignment will follow the Langevin function (section 4.4.4) of ideal superparamagnetism. The rough surface on which the grains settle will result in random deflections of the grain moments. Having settled, the grains may be unable to rotate in response to a change in field, i.e. they have become blocked and are unable to come into equilibrium with a changed field, and the sediment has a remanent magnetization. The equilibrium state before blocking is problematic because of the difficulty in modelling the effect of the surface on which the grains settle. If this is disregarded, the Langevin function, modified to accommodate a range of particle magnetic moments m uniformly distributed from 0 to m_{max}, gives a field dependence of depositional DRM of the form $M = M_0(1/x)\ln((\sinh x)/x)$ where M_0 corresponds to perfect alignment of the grain moments (i.e M_s for monodomain particles) and $x = m_{max}H/kT$. This function is tabulated by Stacey and Banerjee (1974). In the limit of small fields, $M = M_0 m_{max} H/6kT$.

A remanence blocked in the action of settling on the sediment surface may be described as a depositional DRM. Post-depositional DRM in which the grains become blocked only at a later time has also been postulated and has support from laboratory simulations. Grains in water-filled voids may still come into equilibrium with the field before the sediment becomes compacted, or dries out, or the chemical production of cement (itself possibly magnetic) prevents grain rotation. Blocking occurs at some stage during the advance of one or all of these processes. Partial post-depositional DRMs are acquired by each blocking fraction, small grains possibly being the last to lose their freedom to rotate in response to a change in field. The post-depositional processes, including for example the effects of burrowing organisms, are complicated and difficult to model. DRM stands apart from the models for TRM, CRM (and VRM, PRM and ARM, see below) each of which contains a parameter or parameters related to the blocking process (T_B, v_B, $M_s(t_B)$, T_a, H_{pB}). An assemblage of monodomain grains can acquire TRM, CRM, ARM and DRM but whereas the first three depend on the interplay of thermal fluctuations and energy barriers, the last does not. A demagnetized individual multidomain grain can acquire TRM, CRM and ARM but not a DRM, in fields too weak to overcome barriers to magnetization change.

Laboratory studies of depositional and post-depositional DRM indicate that the process can, in principle, account for the stable remanence of many sedimentary rocks, the stability in a mechanically stable sediment being related to the coercive force of the constituent particles. A reversal of the DRM type process, i.e the physical randomization of the grain orientations in the artificial 'rocks' used for laboratory studies of magnetization processes and which are simply pressed compacts or are bound together by soluble material, provides a means of demagnetizing 'hard' systems. Alternatively the induction of a saturation DRM while fabricating artificial samples may provide a high degree of grain alignment when this is required.

6.1.3 *Viscous remanent magnetization*

The role of thermal fluctuations in bringing about 'viscous' magnetization changes has already been discussed in section 4.4.5. A monodomain assemblage acquiring a TRM at T_B is not in equilibrium with the field and temperature on cooling to T_0. If the field remains constant in magnitude and direction, the assemblage will slowly approach a new equilibrium state corresponding to a magnetization $(1/3)\mu_0 M_s^2(T_0)vH/kT_0$, the relaxation time being $(1/f_0)\exp(K_u(T_0)v/kT_0)$. The intensity of magnetization would ultimately be changed by a factor $(M_s(T_0)T_B)/(M_s(T_B)T_0)$, i.e. the remanence as measured will be enhanced compared to the original TRM. A change in field direction will produce a gradual growth of magnetization in the new direction, a viscous remanent magnetization (VRM), and a corresponding decay of the TRM in the original direction. The constituents of an assemblage with a range of v will acquire viscous magnetization at different rates corresponding to their different relaxation times. In an assemblage of identical monodomain grains the stability of the TRM and VRM are identical and no demagnetization process will reveal the direction of the TRM. In an assemblage with a range of heights of barriers to magnetization change, those with the lower barriers will acquire VRM, and lose TRM, faster. The stability of the remanence held by these grains will be lower than a subset of grains with higher barriers and a lower VRM component. The remanence of the system may be progressively destroyed by AF demagnetization leaving a residual remanence which is a closer and closer approximation to the original TRM direction. The peak alternating field required to demagnetize an assemblage of monodomain grains in the absence of thermal fluctuations is $2K_u/\mu_0 M_s$. However, the coercive force may be significantly reduced by thermal fluctuations. In the presence of a field, the barrier to magnetization reversal is reduced, for field and grains aligned, from $K_u v$ (Fig. 4.9(*a*)) to $K_u v(1 - \mu_0 M_s H/2K_u)$ (Fig. 4.9(*c*)) with the corresponding reduction in the time constant. Equating this reduced barrier to $\ln(f_0\tau)kT$ (as in section 4.4.4) we find the relation between applied field and the time constant of magnetization reversal. Having decided on a particular time interval, i.e. the time we are prepared to expose a magnetized grain to a reverse field and wait for reversal to occur, or a typical time associated with an AF demagnetization procedure, the field becomes the coercive force for that experimental time constant. In section 4.4.4 we derived the coercive force for a time constant of 10^2 s, i.e. energy barrier $\sim 25kT$. If we are prepared to wait a longer time, the field necessary for reversal in that time would be lower (Fig. 6.1). In general, the coercive force $H_c = (2K_u/\mu_0 M_s)(1 - (kT\ln(f_0\tau)/K_u v)^{\frac{1}{2}})$ where τ is the time allowed for magnetization reversal to take place. Suppose we have a material with anisotropy energy barriers $K_u v = 50\,kT$—this corresponds to a relaxation time in zero or weak field of 10^{13} s or about 10^6 years. The assemblage may therefore acquire appreciable VRM in the time between reversals of the geomagnetic field. On a time scale of 10^6 years the coercive force of the material is zero. On the laboratory time scale, the coercive force is high at 30% (for $\tau = 10^2$ s) of the coercive force in the absence of thermal fluctuations (i.e. $\tau \sim 1/f_0$). This may exceed the field available from AF apparatus in the case of particularly 'hard' minerals, and this procedure would fail to reveal the less distorted TRM of a more stable fraction with $K_u v > 50\,kT$.

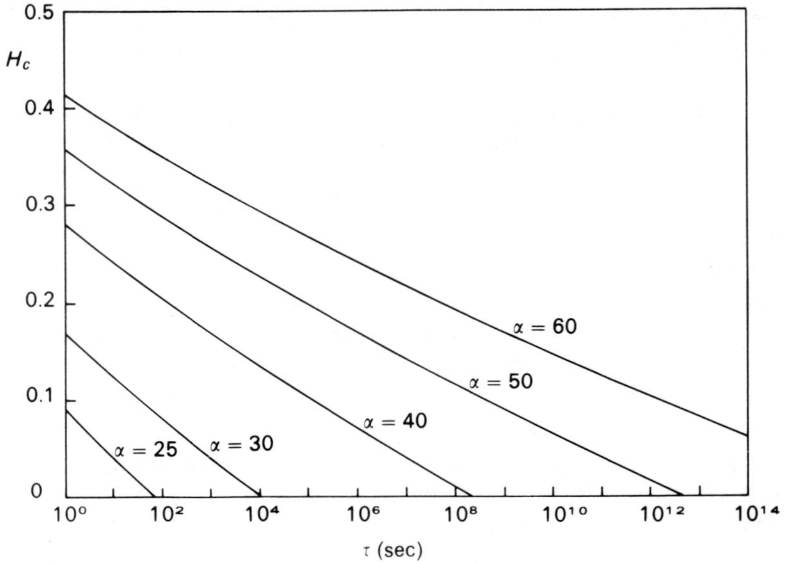

Figure 6.1 The stability of TRM or CRM over geological time. The 'coercive force' is not a fixed switching field but depends on the time available for reversal. Over the time available in nature, coercive force may be small or zero. Here the time dependence of coercive force is shown (in units of H^*) for an aligned assemblage of identical grains with uniaxial anisotropy barriers to magnetization reversal $K_u v = \alpha kT$. H^* is the coercive force in the absence of thermal fluctuation, $2K_u/\mu_0 M_s$. The parameter for the curves is volume, expressed in units of kT/K_u.

As mentioned previously, although a simple system would gain or lose remanence according to an exponential time dependence, such a form is rarely observed in practice. The commonly observed logarithmic time dependence has been attributed in the case of monodomain grains to a particular relaxation time spectrum or to the effect of inter-particle interactions.

6.1.4 *Piezoremanent magnetization*

The application of stress to a magnetic material introduces an induced anisotropy via magnetostriction (section 3.5) and thus will combine with other anisotropies to form the barriers to magnetization reversal in monodomain grains. The barriers will be different in the presence or absence of stress. The effect of stress may be to bring barriers, which otherwise would be too high, into the range of thermal fluctuations. Equilibrium with an ambient field would then be reached. The removal of stress restores the barriers to their unstressed height and the magnetization (PRM) is blocked in. Similarly, in multidomain particles, stress modifies domain wall barriers, and domain wall movements, which would otherwise take place very slowly, bring the grain into equilibrium with a field. In the absence of a field, pressure demagnetization may occur under the action of a stress.

6.2 Mechanisms in the laboratory

The division into mechanisms in nature and in the laboratory is to some extent artificial, as all processes can be simulated, although often as highly imperfect analogues to the natural process, in the laboratory. The mechanisms described in this section are not believed to contribute to the palaeomagnetic record.

Isothermal remanent magnetization (IRM) is acquired most often in the course of laboratory experiments, but can be acquired in nature on exposure to a field large enough to overcome the barriers opposing magnetization change, and the field removed. This occurs in the vicinity of lightning strikes. Soft material may acquire IRM in the fields due to manmade objects, during sampling or while in transit to the laboratory. The acquisition of IRM and more specifically saturation IRM, M_{rs}, is dealt with in sections 4.3.1 to 4.3.3.

6.2.1 *Anhysteretic remanent magnetization*

The process of AF demagnetization, as described in section 4.3.4, represents a fixed switching field model, i.e. the magnetization may be reversed only when a specific field is applied. This model is strictly valid only at the absolute zero of temperature or if the time scale of the experiment (a half period of the alternating field) is of the order of $1/f_0$. As we have seen from our discussion of the time dependence of magnetization and of the role of thermal agitation generally in the remanence impressing mechanisms, any finite field will cause reversal of magnetization. The lower the field the longer it must be applied for a significant effect. In the random assemblage of Figs. 4.13 and 4.14, each annular shell of the system may be demagnetized by an alternating field with amplitude lower than the corresponding field in the figures—the nearer the field approaches this value the more rapidly will the shell be demagnetized. A field lower than $K_u/\mu_0 M_s$ will demagnetize the whole assemblage, each elementary shell requiring a progressively longer application as ϕ departs from 45°.

Consider an aligned monodomain assemblage subject to an alternating field of amplitude $H_p(< 2K_u/\mu_0 M_s)$ the assemblage having previously been magnetized to saturation, although the final outcome is the same regardless of the initial state. The state of the system during the first opposed half cycle of the alternating field is sketched in Fig. 6.2(*a*). The barrier to magnetization reversal, i.e. rotation into the field direction of that half cycle is reduced to $K_u v(1 - H_p/H^*)^2$ (where we have replaced $2K_u/\mu_0 M_s$ by H^*) which, when equated to $\ln(f_0\tau).kT$, gives the corresponding relaxation time τ. The population of the elevated level of Fig. 6.2(*a*) decays exponentially with time, or the rate of decay of the population is $dN/dt = -N/\tau$ where N is the population at time t. The barrier opposing the return, to the original direction, of the grains which have reversed is large, and we may neglect any flow of population in this direction during this half-cycle. The total flow out of the original

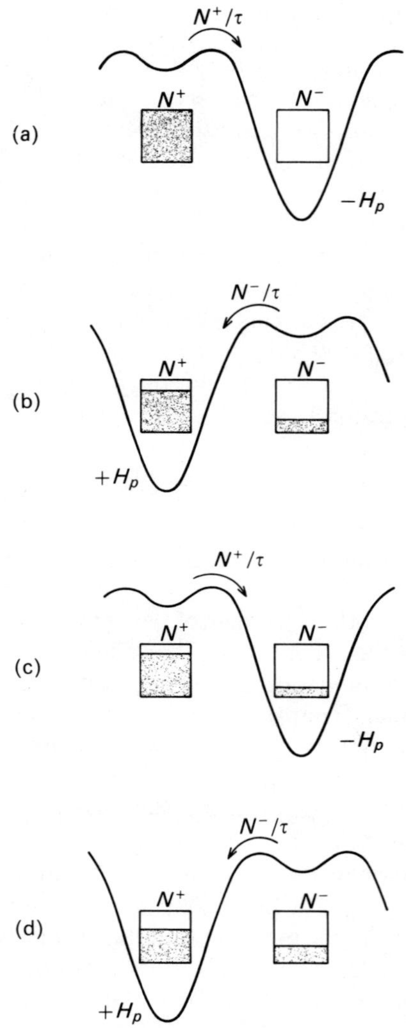

Figure 6.2 The application of an alternating field H_p with amplitude less than $2K_u/\mu_0 M_s$ to an aligned assemblage of monodomain grains initially magnetized to saturation in the + direction. (*a*) shows the first opposed half-cycle of H_p, during which a part of the population transfers to the − direction, the rate depending on the time constant τ and the instantaneous population of the + direction N^+. Reverse flow may be ignored, in this and all subsequent stages. (*b*) The second half cycle in which some grains return to the + direction, the time constant being the same, but the flow smaller because N^- is smaller than the original N^+. (*c*) The third half-cycle in which the + → − migration rate is smaller than it was in (*a*) because N^+ is now smaller. (*d*) The − → + migration is greater than (*b*) because N^- is now greater. N^+ decreases and N^- grows until the two populations are on average the same and gradual reduction of H_p leaves blocked equal + and − populations. In the acquisition of ARM, τ^+ and τ^- must be distinguished and the blocked N^+ and N^- are no longer the same.

direction is found by integrating the rate over the half-cycle, which is simplified by adopting a square wave periodic field which maintains τ constant. In the next half-cycle (Fig. 6.2(*b*)) the relaxation time for the return to the original direction is the same reduced τ of the first half-cycle, but the rate of flow, depending as it does also on the instantaneous population (i.e. $-N/\tau$), is much smaller and the transfer during this half-cycle, much less. In the third half-cycle (Fig. 6.2(*c*)) the flow, out of the original direction, is again governed by the same τ but is less than half-cycle 1 because of the slightly reduced population. The return flow of half-cycle 4 into the original direction is greater than the corresponding return flow of half-cycle 2 because the population is now greater. The pattern is an initially large but decreasing transfer per cycle out of the original direction, and an initially small but growing transfer back to the original orientation. Eventually the two rates become the same and a fraction of the whole population is transferred back and forth between the two equally populated (on average) orientations. If the field were now instantaneously removed, an excess of magnetization would be left in the direction of the last application of H_p. Reducing H_p to a lower value results in a new dynamic equilibrium where the fraction undergoing continuous transfer is smaller, and the instantaneous net imbalance is also smaller. Gradual reduction of H_p leads to a near perfect cancellation of magnetization. Because the process involves thermal fluctuations, there is no field below which magnetization change does not take place but, once again, an unblocking point H_{pB} can be conventionally defined because τ is a strong function of field. We may note that an aligned assemblage of identical monodomain grains cannot be demagnetized according to a fixed switching field model but requires thermal fluctuations for effective AF demagnetization.

It appears that magnetization and demagnetization processes exist in complementary pairs, the perturbation process which brings magnetization change into the range of thermal fluctuations taking place in the presence or absence of a field in either case. During AF demagnetization the applied periodic field allows the effective action of thermal fluctuations. This leads to a net remanence in the presence of a weak constant field, which would otherwise produce no such remanence, when the two fields are applied simultaneously. We now have the superposition of the periodic field H_p and the quiescent field H_q so that during (say) the positive half-cycles the resultant field is $H_p + H_q$ and in the negative half cycles $H_p - H_q$. The time constants for flow into the direction of H_p during the positive and negative half-cycles are no longer the same. $1/\tau^+ = f_0 \exp\{-K_u v(1 - H_p/H^* - H_q/H^*)^2/kT\}$ and $1/\tau^- = f_0 \exp\{-K_u v(1 - H_p/H^* + H_q/H^*)^2/kT\}$ so that $1/\tau^+ < 1/\tau^-$, and flow rates into the direction corresponding to H_p and H_q parallel will tend to be higher than the opposite direction until equilibrium is reached. The sequence of events follows the same pattern as Fig. 6.1 for an initially magnetized assemblage and an equivalent picture can be devised for an assemblage which has the two directions initially equally populated. The final dynamic equilibrium is the same in either case with an equal transfer of population between the two directions in each cycle. Thus the average values of dN^+/dt and dN^-/dt are the same where N^+ and N^- are the average populations of the directions parallel and antiparallel to H_q. $dN^+/dt = -N^+/\tau^-$ where τ^- is the time constant for flow out of the plus direction (H_p and H_q opposed). Thus $N^+/N^- = \tau^-/\tau^+ > 1$ and the average excess magnetization in the H_q direction is $M_s(N^+ - N^-)/(N^+ + N^-)$. If the amplitude of the periodic field now falls, the fraction transferred back and forth becomes negligibly small and the system has a net static excess population in the H_q direction, the actual frozen-in population depending via τ^+ and τ^- on the value of H_p at the blocking point. The resultant anhysteretic remanent magnetization (ARM) becomes $M_s \tanh\{(\mu_0 M_s v H_q/kT)(1 - H_{pB}/H^*)\}$ with the usual corresponding weak-field form (Jaep, 1971).

A randomly orientated assemblage requires numerical solution as in the case of TRM and CRM. In the weak field limit, the model ARM would be $(1/3)\mu_0 M_s^2 v(1 - H_{pB}/H^*)H_q/kT$. In the equivalent TRM model (section 5.2.2) where the field also lies at all angles ϕ with respect to the dispersed unique axes of the grains, the simplifying assumption, made prior to numerical integration, is that the moments are not deflected by the field from the unique grain axes, i.e. the energy barrier to magnetization reversal, and therefore blocking temperature, is independent of orientation. The equivalent simplification in the ARM model is that the ratio H_{pB}/H^* applies for all orientations. The angular dependence of the coercive force in the absence of thermal fluctuations, or fixed switching field, is shown in Fig. 4.13 and the effect on barriers to magnetization change of a field applied at 45° to the unique axis shown in Fig. 4.9 (f). It seems likely that the angular dependence of the blocking condition is significant and that each annular cone will make different proportional contributions to a TRM or ARM, and the AF demagnetization characteristics of the two types of remanence in the same system would differ. The angular dependence of ARM blocking should be reduced by 'tumbling' the specimen in the alternating field so that each grain may have the same blocking condition (Stephenson, 1983). The ratio of TRM to ARM with angle-independent blocking in both cases and for weak fields is $(M_s(T_B).T_0)/(M_s(T_0).T_B(1 - H_{pB}/H^*))$.

The fixed switching field model of AF demagnetization also predicts ARM acquisition. However, this would predict no remanence when $H_p + H_q < K_u/\mu_0 M_s$ and saturation of that part of the assemblage with switching fields $< (H_p + H_q)$ as soon as H_q was bigger than the incremental fall per cycle of H_p, i.e. normally for microscopically small H_q. The thermal fluctuation model is needed to successfully describe the observed characteristics.

A multidomain ARM model has the domain wall passing backwards and forwards across the grain with progressively smaller amplitude until it settles at the equilibrium point where $H = -NM$. The model TRM/ARM ratio in the same grain is then simply $M_s(T_0)/M_s(T_B)$.

Partial ARM may be defined by analogy with the equivalent partial TRM, CRM, etc., when H_q is applied for only part of the range of H_p. Total ARM is induced when H_p exceeds the maximum coercive force (corresponding to the experimental time) of the constituents of the assemblage. Saturation ARM will occur in a similar way to saturation TRM and the ARM intensity will depend, via the blocking condition, on the rate of reduction of H_p.

6.2.2 *Gyroremanent magnetization and rotational remanent magnetization*

Figure 4.9 shows irreversible magnetization change as an inert object sliding down a slope into a potential well. The magnetization vector makes a transition from $\theta = 0$ to $\theta = 180°$ in Fig. 4.9, in which the field is applied along the 180° direction. The intermediate values of θ, through which the vector passes on making the irreversible flip, do not define unique orientations but only some position on an annular cone. We might imagine that the transition is made by the most direct route, the tip of the magnetization vector describing a circular arc. This is to ignore the fact that the magnetization vector has intrinsic angular momentum and that the application of a torque, such as is represented by the gradient of the energy curves of Fig. 4.9, results in precession. In the absence of damping, i.e. irreversible energy transfer between the magnetization vector and the crystal lattice, no irreversible flip of the magnetization would take place. A constant θ would be maintained, or the inert object of Fig. 4.9 would perch on the sloping energy profile without falling. This phenomenon is easily demonstrated by placing a spinning top on an incline down which it would slide if it were not spinning. The spinning top migrates sideways across the slope, the direction of migration depending on the direction in which it has been set spinning.

Damping causes the vector to spiral into the minimum energy orientation, light damping allowing many precessional rotations, heavy damping none. The intrinsic angular momentum remains unchanged; the energy given up to the crystal is the potential energy the vector gained by application of the reverse field. The path traced out by the vector during the irreversible rotation is central to theories of gyroremanence and rotational remanence as they presently exist, such theories still being in the developmental stage.

Gyroremanence (GRM) is the remanence acquired in an anisotropic (aligned) assemblage of monodomain grains to which an alternating field is applied at an angle to the alignment axis. The existence of GRM was postulated by A. Stephenson who also demonstrated its occurrence in artificial and natural anisotropic assemblages (Stephenson, 1981). The GRM lies at right angles to the AF axis and the alignment axis. The application of an AF field at angle ϕ to the unique axis of a monodomain grain which could be

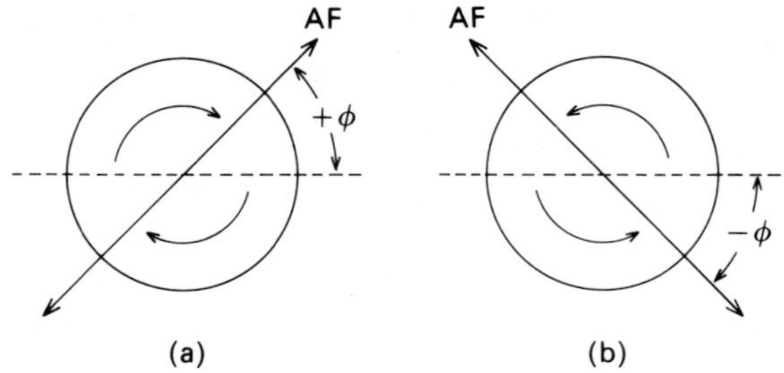

Figure 6.3 In the application of an alternating field at angle ϕ to an aligned assemblage of uniaxial grains, the sense of rotation of the magnetization vector is different for $+\phi$ and $-\phi$. Because of the dynamics of magnetization change, the magnetization vector makes an excursion out of the plane containing the field and anisotropy axes, in the case of $+\phi$ above the plane of the diagram, and for $-\phi$ below the plane of the diagram. Given some suitable and more complicated grain anisotropy, a remanence (GRM) may be blocked at right angles to the AF and the assemblage axes either 'up' or 'down' depending on the sign of ϕ.

part of an aligned assemblage is shown in Fig. 6.3(*a*). The magnetization vector flips backwards and forwards in response to the alternating field. Because of dynamics of the magnetization reversal, the rotating vector does not remain in the plane of the diagram but is deflected out of the figure during reversal. This is so for reversals in both directions. An AF field applied at an angle of $-\phi$ (Fig. 6.3(*b*)) produces reversals which are accompanied by deflections into the figure. In those simple ideal aligned systems any remanence can lie only in one direction or another along the preferred direction. The acquisition of GRM, which is a remanence at right angles to both the preferred axis and the axis of the AF field, requires a more complicated system. The precise form of the grain anisotropy which will account for GRM, and the role of thermal fluctuations have still to be determined, but the change in sign of induced GRM with angle between AF axis and sample axis has been convincingly demonstrated by experiment, and the remanence would seem to be linked with the path taken by the magnetization vector during irreversible rotation.

In an assemblage with randomly dispersed grain axes, the grains present all ϕ, both positive and negative, to the AF field, excursions will take place in all directions uniformly and any GRM acquired will be internally cancelled within the assemblage. This is no longer the case when the assemblage is rotated (Stephenson, 1980) as can be seen from Fig. 6.4. In the first half-cycle of the alternating field $+H_p$ magnetizes the assemblage along a particular direction AB and, on dying away to zero, leaves a remanence in that direction. This now defines a specific direction in the isotropic assemblage. If the rotation rate is zero, the field moving to $-H_p$ in the second half-cycle reverses the magnetization (the range of switching field producing a progressive reversal as H_p rises). The irreversible flips occur in all directions without favouring either sense of reversal. If the assemblage is rotated anticlockwise at a frequency lower than the AF frequency (Fig. 6.4(*c*)) the line AB has now

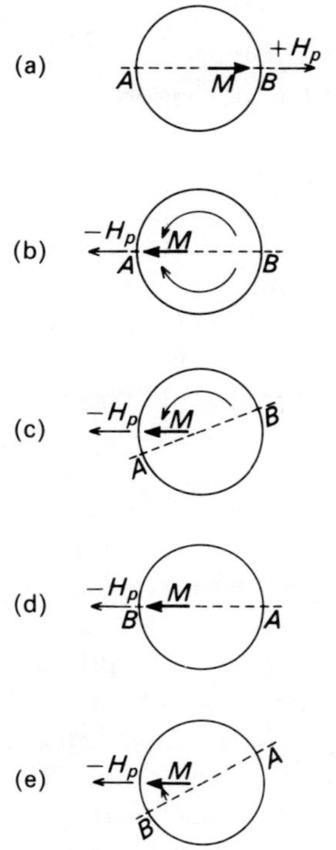

Figure 6.4 The effect of rotation on the sense of magnetization reversal by an alternating field. (*a*) A positive half-cycle of the AF magnetizes the system and defines a direction AB in the randomly orientated assemblage. (*b*) Without rotation $-H_p$ produces reversal without any predominant sense of irreversible magnetization rotation. (*c*) Anticlockwise rotation at frequency lower than the AF frequency produces an anticlockwise magnetization reversal with excursions below the plane of the diagram and GRM antiparallel to the rotation vector. (*d*) No magnetization reversal occurs when the rotation and AF frequencies are the same. (*e*) Rotation frequency greater than the AF frequency produces clockwise magnetization reversal, 'upward' excursion and GRM parallel to rotation vector.

advanced towards the $-H_p$ direction and irreversible rotation when H_p is large enough takes place in the anticlockwise direction only. The changed magnetization now defines a new AB axis and the same sequence is repeated every half-cycle. As rotation rate increases, the flip will always be anticlockwise provided the rotation frequency is less than the AF frequency. The corresponding damped precessional excursion is downwards into the plane of the diagram and a trapped GRM (termed an RRM in this context) therefore lies antiparallel to the rotation vector. When the rotation frequency and AF frequency are the same (Fig. 6.4(*d*)), no magnetization reversals take place, the line AB always experiencing H_p in the same direction. At rotation frequencies greater than the AF frequency, the line AB has now passed the $-H_p$ direction when the maximum negative reverse field occurs (Fig. 6.4(*e*)), and the magnetization rotates in the anticlockwise direction with the corresponding excursion out of the plane of the figure. Again this will be repeated each half-cycle with a newly defined *AB* direction. The trapped GRM will now be parallel to the rotation vector.

6.3 Summary

In the CRM, PRM and ARM mechanisms, the perturbation appropriate to the process—small volume, pressure, alternating field—reduces the barriers to magnetization change and brings them into the range of thermal fluctuations so that equilibrium may be reached with an applied field. Removal of the perturbation blocks in the equilibrium state as a remanent magnetization. On a given time scale the remanence is permanent, and a new equilibrium with a changed applied field cannot be approached. VRM is accompained by no special 'perturbation' but simply corresponds to exposure to a changed field for a long enough time that the 'given time scale' on which TRM, CRM, etc., are stable has been exceeded. The stability of a remanence or its coercive force are not absolutes but depend on the time available for reversal of magnetization to occur. Because of the difference in the times available in nature and in the laboratory, a TRM or other remanence may have zero coercivity in nature and relatively high coercivity in the laboratory. The stabilities to AF demagnetization of TRM, CRM and VRM in the same assemblage of identical grains are identical. The stability of ARM may not be the same as TRM, etc., in a randomly aligned assemblage of monodomain grains, because the angular dependence of the blocking condition is not the same, although this depends on the experimental ARM-inducing procedure.

DRM stands apart from the other remanence-inducing mechanisms in that thermal fluctuations (except insofar as Brownian motion may agitate a suspended particle) do not play a role, and the process is difficult to model satisfactorily.

Models for GRM and RRM are still in a developmental stage, but the correlation between the damped precessional motions of magnetization vectors while in the process of reversing and the observed magnetization implanted in stationary anisotropic or rotating isotropic assemblages by AF, seem to indicate that the dynamics of magnetization reversal are at the root of GRM and RRM acquisition.

Selected bibliography

*Collinson, D.W. (1965) Depositional remanent magnetization in sediments. *J. Geophys. Res.* **70**, 4663–4668.

*Dunlop, D.J. and Stirling, J.M. (1977) 'Hard' viscous remanent magnetization (VRM) in fine-grained hematite. *Geophys. Res. Lett.* **4**, 163–166.

*Fuller, M.D. (1970) Geophysical aspects of palaeomagnetism. *CRC Critical Revs. Solid State Sci.* 137–219.

Jaep, W.F. (1971) Role of interactions in magnetic tapes. *J. App. Phys.* **42**, 2790–2794.

Stacey, F.D. and Banerjee, S.K. (1974) *The Physical Principles of Rock Magnetism*. Elsevier, Amsterdam.
Stephenson, A. (1980) Rotational remanent magnetization and the torque exerted on a rotating rock in an alternating field. *Geophys. J.R. astr. Soc.* **62**, 113–132.
Stephenson, A. (1981) Gyroremanent magnetization in a weakly anisotropic rock sample. *Phys. Earth Planet. Int.* **25**, 163–166.
Stephenson, A. (1983) Changes in direction of remanence of rocks produced by stationary alternating field demagnetization. *Geophys. J.R. astr. Soc.* **73**, 213–239.
*Tucker, P. (1980) A grain mobility model of post-depositional realignment. *Geophys. J.R. astr. Soc.* **63**, 149–163.

Appendix to chapters 4-6: Summary of the magnetization process and weak field remanence acquisition mechanisms.

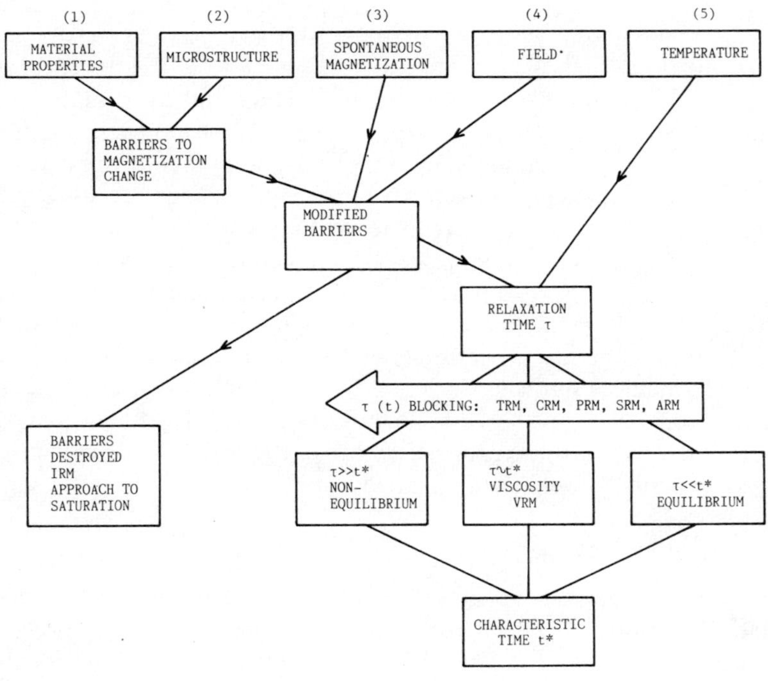

The magnetization process and weak field remanence acquisition mechanisms. A suitable time dependence of one or more of (1) to (5) constitutes a remanence-inducing "perturbation". The complementary demagnetization processes correspond to the same perturbations in the absence of a weak steady field.

(*sources not specifically cited in the text.)

7 Magnetic properties of titanomagnetites and titanomaghemites

7.1 The titanomagnetites

Chapters 3 to 6 have been largely concerned with models for the magnetic behaviour of materials. Chapter 3 dealt with intrinsic properties—those determined by the species of atoms or ions which make up the material, and their interaction with one another in the crystalline environment. Models for the magnetization process (Chapters 4 to 6), i.e. the effect of an applied field, rely on non-intrinsic microstructure, linked to material properties. Chapters 5 to 6 dealt with a special aspect of the model magnetization process in which thermal activation allowed a weak field to impress a relatively strong remanent (non-equilibrium) magnetization, or, in zero field, brought about the removal of a remanence. In this and the following chapter we enter the real world by considering the actual observed properties of magnetic minerals, or assemblages, or intergrowths of minerals, such as occur in the crust of the Earth. Whereas some of the data has indeed been obtained from minerals, i.e. naturally occurring examples of the compounds of interest, much of our knowledge of the magnetic properties of the remanence carriers of rocks is based on laboratory-prepared synthetic analogues. We deal first with intrinsic properties, then with the extrinsic magnetization process and, finally, weak field remanences.

7.1.1 *Intrinsic properties*

Intrinsic properties depend on composition and first we must specify the composition range to which the term 'titanomagnetite' is to apply. The composition parameters x and z (and R) were introduced in Chapter 2. We may arbitrarily define the composition zone for a simple (i.e. unsubstituted) titanomagnetite $Fe_{(3-x)R}Ti_{xR}\square_{3(1-R)}O_4$ as lying in the range $0 < x < 1, 0 < z < 0.1$ $(1 > R > 8/(8.1 + 0.1x))$, any Fe-Ti spinel oxide outside this zone being 'titanomaghemite'. A complex titanomagnetite containing other divalent and trivalent species than Fe^{2+} and Fe^{3+} may be arbitrarily confined to a zone in which the non Fe-Ti cations constitute less than 10% of the total, i.e. the $(a + b + c)$ of section 2.1.1 is less than 3/10. A composition outside this zone would then be a member of a complicated solid solution between a particular titanomagnetite and some multicomponent solid solution system which has special compositions $MgAl_2O_4$ (spinel), $FeAl_2O_4$ (hercynite), $MgFe_2O_4$ (magnesio-ferrite), Mg_2TiO_4 (magnesian-titanite), $MnFe_2O_4$ (jacobsite) and other compounds including chromite ($FeCr_2O_4$) and picrochromite ($MgCr_2O_4$). The same restrictive definition may be applied to titanomaghemites, compounds lying outside the defined zone in a, b, c, x, z space being non-stoichiometric oxidation products of the members of the complex multicomponent solid solutions.

In understanding the intrinsic magnetic properties of any spinel compound, the distribution of cations between the octahedral and tetrahedral sites of the structure is of primary importance. On the assumption of a simple Néel *A-B* ferrite (section 3.6.3) and known ionic magnetic moments, the low temperature saturation magnetization has a simple relation with, and is sensitive to, cation distribution. Other intrinsic properties have a lower sensitivity or a more complicated relationship to the distribution. A common procedure is to infer a cation distribution from saturation magnetization measurements and use this as a basis for the understanding of the observed Curie point temperature, thermomagnetic curves, anisotropy and magnetostriction. Saturation magnetization measurements provide a unique cation distribution when only two cationic species (including vacancies) are present. The presence of more than two species requires the measurement of more than one physical property which has a useful model relationship to the distribution. Among such properties are the intensities of X-ray and neutron diffraction lines, electrical conductivity, Mössbauer effect, optical and infra-red spectroscopy, and ferrimagnetic resonance. Simple titanomagnetites contain four species, $Fe^{2+}, Fe^{3+}, Ti^{4+}$ and $\square$. In practice, investigators of what might be termed 'near-stoichiometric' titanomagnetites or materials prepared at high temperature at which only a limited degree of non-stoichiometry can be supported (Fig. 2.11), generally neglect the presence of vacancies. High temperature non-stoichiometry has a large, but unexplained, effect on Curie temperature (section 7.1.1.2) but otherwise the significance of the error introduced by neglect of high temperature vacancies into discussions of the intrinsic properties is uncertain.

7.1.1.1 *Saturation magnetization.* The cation distribution for a simple titanomagnetite may be written:

$$Fe^{2+}_{\delta}Fe^{3+}_{1-\delta}|Fe^{3+}_{(2-2x+z(1+x))R-1+\delta}Fe^{2+}_{(1+x)(1-x)R-\delta}Ti^{4+}_{xR}\square_{3(1-R)}|O_4$$

where octahedral site occupants (the magnetic *B*-sublattice) are within the parentheses and the tetrahedral site occupants (the *A*-sublattice) are outside. That Ti^{4+} is located only on octahedral sites is a widely accepted working assumption. The assignment of high temperature vacancies to octahedral sites has also been assumed in the above distribution. δ is an adjustable parameter which can be chosen to fit the distribution to the

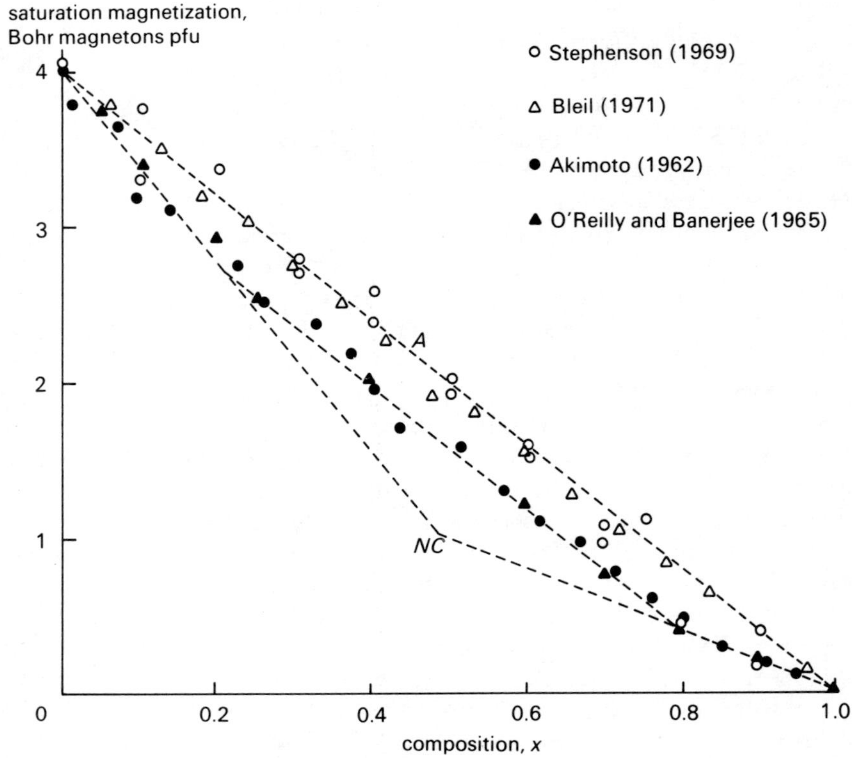

Figure 7.1 Saturation magnetization data (in β pfu) for the titanomagnetite solid solution, obtained mainly from sintered material of unknown degree of non-stoichiometry. The dashed lines represent the saturation magnetization according to various models for the Fe^{2+}–Fe^{3+} distribution in stoichiometric titanomagnetite (see text).

observed saturation magnetization, δ being a function of x and z. Taking spin-only moments (5β for Fe^{3+} and 4β for Fe^{2+}) and the *A-B* collinear spin structure, the saturation magnetization in β per formula unit (pfu) is $n_\beta = (14 - 6x + z + zx)R + 2\delta - 10$.

A selection of saturation magnetization data is given in Fig. 7.1. For the purposes of determining cation distribution, the measurements are ideally carried out at absolute zero and in infinite field. The experimental conditions employed in the various investigations present varying degrees of approximation to this ideal. The experimental data shows some variation, and this led Stephenson (1969) to suggest that the distribution was temperature-dependent. Pre-existing cation distribution models—the Néel/Chevallier model (*NC* in the figure) in which the site preference of Fe^{3+} for tetrahedral sites dominates (thus $\delta = 0$, $x < 0.5$; $\delta = 2x - 1$, $x > 0.5$) and the Akimoto model ($\delta = x$) (*A* in the figure)—were reinterpreted as corresponding to a low temperature, ordered, $Fe^{2+} - Fe^{3+}$ distribution and a high temperature, disordered, $Fe^{2+} - Fe^{3+}$ distribution, respectively. The observed data, tending to lie between NC and A in Fig. 7.1 would then correspond to the equilibrium distribution at the finite temperature from which the samples were rapidly cooled. A related system $Mg_{1+x}Fe^{3+}_{2-2x}Ti_xO_4$ $(0 < x < 1)$, which may be regarded as titanomagnetite with the Fe^{2+} replaced by Mg^{2+}, does indeed exhibit a temperature-dependent cation distribution, samples quenched from high temperature ($>1000°C$) having higher saturation magnetization than those which are slowly cooled over a period of several hours (Blasse, 1964). The quenched samples may have a near-random disordered distribution of Mg^{2+} and Fe^{3+} between tetrahedral sites and those octahedral sites not occupied by the Ti^{4+}, which has a strong octahedral site preference and may not be subject to the temperature-dependent order/disorder of the other cations. Similar quenching and annealing experiments on titanomagnetite fail to reveal a temperature-dependent distribution even when Mg^{2+} is also incorporated into the system (O'Donovan and O'Reilly, 1980). The Fe^{3+}–Mg^{2+} interchange observed in other systems, such as that mentioned above and $Mg^{2+}Fe^{3+}_2O_4$, appears to be suppressed by the presence of Fe^{2+} ions. This may be because Fe^{2+} and Fe^{3+} do not coexist as individual entities in the spinel lattice at room temperature and above. The rapid exchange of the extra electron of the Fe^{2+} results in averaged physical properties of the Fe ions within a sublattice of the structure. In effect, the lifetime of an individual Fe^{3+} is short compared to the characteristic time of Fe^{3+}–Mg^{2+} interchange. Intra-sublattice electron hopping may therefore result in no measurable temperature-dependence of cation distribution in the titanomagnetites or Mg-substituted counterparts.

An alternative explanation for the apparent absence of the temperature-dependent distribution in titanomagnetites is that there is indeed a dependence of the Fe^{2+}–Fe^{3+} distribution on temperature, but that measurements at room temperature only reveal the distribution appropriate to room temperature. The distribution corresponding to some temperature from which the sample has been quenched cannot be preserved at room temperature because inter-sublattice electron hopping rapidly brings about the equilibrium Fe^{2+}–Fe^{3+} distribution (Jensen and Shive, 1976). The variation in saturation magnetization for a given x value is still to be explained, however. One possibility is the undetermined variations of non-stoichiometry in the materials of Fig. 7.1. A vacancy concentration of up to about 0.05 pfu can be accommodated in the region of high-temperature non-stoichiometry. The assumption that such vacancies are confined only to octahedral sites needs to be justified and interchange of vacancies with (say) Fe^{3+} would produce an enhanced saturation magnetization. Finally it will be observed from Fig. 2.11 that the maximum permitted vacancy concentration in the high-temperature monophasic region is itself temperature-dependent.

A typical range for the saturation magnetization of $Fe_{2.4}Ti_{0.6}O_4$ at room temperature is 20–30 Am^2/kg ((1–1.5) × 10^5 A/m) and a typical value for Fe_3O_4 about 92 Am^2/kg (4.8 × 10^5 A/m). The saturation magnetization of the series at room temperature can be approximated to 92(1–1.25x) Am^2/kg. The saturation magnetization falls approximately linearly with concentration of substituting diamagnetic ion, 0.1 Mg^{2+} pfu reducing the magnetization of $Fe_{2.4}Ti_{0.6}O_4$ by about 15% and 0.1 Al^{3+} pfu reducing the magnetization by about 25%.

The variation of saturation magnetization with temperature in Fe_3O_4 is that of a *Q*-type ferrite (Fig. 3.16) whereas for $x > 0.6$ a *P*-type curve appears. A variety of mechanisms have been proposed for this behaviour, some involving exchange interactions—a positive *BB* interaction (Schult, 1970) or a composite molecular field representing the *AB* interaction between Fe^{2+} and Fe^{3+} on the two sublattices (Stephenson, 1972)—others special properties of Fe^{2+} ion (Banerjee *et al.*, 1967; Ono *et al.*, 1968). The variation of M_s with temperature for $x = 0.4$ and $x = 0.6$ is tabulated by Özdemir and O'Reilly (1981).

7.1.1.2 *The Curie temperature.* The Curie point temperature falls approximately linearly with composition from 575°C ($x = 0$) to −153°C ($x = 1$) and is readily explicable in terms of the dilution of the *AB* interaction by non-magnetic Ti^{4+} and replacement of the strongly interacting Fe^{3+} by weakly interacting Fe^{2+}. As mentioned in section 3.3, according to the molecular field model, the Curie temperature is more strongly dependent on the bulk composition of a ferrite than on the cation distribution. Thus the adapted molecular field model of Stephenson (1972) (section 3.3), by by which the Curie temperatures of members of the solid solution can be calculated relative to the observed values for the end members, suggests that observed variations of Curie point for a given *x* are too large to be explicable in terms of variations in Fe^{2+}–Fe^{3+} distributions. It seems most likely that non-stoichiometry is a much more important influence. Hauptman's (1974) study of the effect of high temperature non-stoichiometry on the Curie point temperature of $Fe_{2.4}Ti_{0.6}O_4$ and $Fe_{2.6}Ti_{0.4}O_4$ may be summarized by the relation $T_c(x, z) = T_c(x, 0) + 700z$ where $T_c(x, 0)$ refers to exact stoichiometry. For $Fe_{2.4}Ti_{0.6}O_4$ ($x = 0.6$), a good value for $T_c(x, 0)$ seems to be 150°C. Taking the maximum non-stoichiometry for the monophasic region to be given by $z \sim 0.1$, the Curie point for '$Fe_{2.4}Ti_{0.6}O_4$' can range from 150°C to 220°C and such a variation can be found in the literature. It is not clear that molecular field theory can account for the high Curie temperatures of the non-stoichiometric compositions unless the strength of the *AB* interaction can somehow be enhanced by defects introduced at high temperatures at the atomic level. In any event this seems to be an important experimental result which must also have implications for the Curie temperatures of titanomaghemites obtained from the titanomagnetites by oxidation.

As might be expected, a reduction in Curie temperature follows from substitution of Fe ions by diamagnetic species. 0.1 Al^{3+} pfu (replacing Fe^{3+}) reduces the Curie point of $Fe_{2.4}Ti_{0.6}O_4$ by about 50°C whereas 0.1 Mg^{2+} pfu (replacing Fe^{2+}) produces a much smaller reduction.

The increase in Curie temperature with applied hydrostatic pressure can be understood in terms of an actual reduction in metal-oxygen distances and enhancement of the superexchange interaction. The rate of increase ($\sim 10^{-8}$K/Pa) (Schult, 1970) may not significantly affect calculations of the Curie temperature isotherm in the Earth's crust.

7.1.1.3 *Magnetocrystalline anisotropy and magnetostriction.* The cubic magnetocrystalline anisotropy constants K_1 and K_2 have been determined for single crystal specimens of the titanomagnetites by torque magnetometry. The principal features of the data are compensation points—temperatures at which K_1 is zero—and the large anisotropy introduced in compositions with high x. The observed compensation points occur below room temperature, although compositions in the range $0.65 \lesssim x \lesssim 0.8$ may have compensation points between room temperature and the Curie point. Because the Curie temperatures of the high x values are near (or below) room temperature, the very large anisotropy of such compositions will produce a 'hard' magnetism mainly below room temperature and have little direct relevance in the geophysical context. The zero in K_1 in magnetite occurs at about 130°K. Below this temperature, K_1 is positive (and $\langle 100 \rangle$ the preferred directions) and above, negative ($\langle 111 \rangle$ directions preferred). The negative K_1 increases as temperature rises and reaches a maximum negative value of about -1.5×10^4 J/m^3 at about 220°K and thereafter falls steadily. The variation of K_1 for magnetite above room temperature (Fletcher and O'Reilly, 1974) is shown in Fig. 7.2. Similar behaviour is observed in compositions $x = 0.04$ and $x = 0.1$ and, in this temperature range, the temperature-dependence of K_1 for all three compositions can be expressed in the form $K_1(T) \propto M_s^p(T)$ where $p \sim 8$.

Below room temperature, compositions in the range $0 < x < 0.3$ show a similar K_1 temperature dependence to $x = 0$, the zero point falling with increasing x to temperatures below 80°K. For $x = 0.04$ and $x = 0.1$ the zero points in K_1 occur at 108°K and 90°K respectively (Syono, 1965). More titanium-rich compositions ($x = 0.56$, $x = 0.68$) have large positive values of K_1 at 80°K ($\sim +10^6$ J/m^3) and fall rapidly with increasing temperature, $x = 0.56$ having a zero point in K_1 at 230°K. For the studied compositions, the room temperature values of K_1 are: $x = 0$, $K_1 = -1.4 \times 10^4$ J/m^3; $x = 0.04$, $K_1 = -1.9 \times 10^4$ J/m^3; $x = 0.1$, $K_1 = -2.5 \times 10^4$ J/m^3; $x = 0.18$, $K_1 = -1.9 \times 10^4$ J/m^3; $x = 0.31$, $K_1 = -1.8 \times 10^4$ J/m^3; $x = 0.56$, $K_1 = -7 \times 10^3$ J/m^3; $x = 0.68$, $K_1 = +1.8 \times 10^3$ J/m^3 (Syono, 1965). Thus the variations in K_1 with composition at room temperature are probably not great for most of the composition range. Somewhere between $x = 0.56$

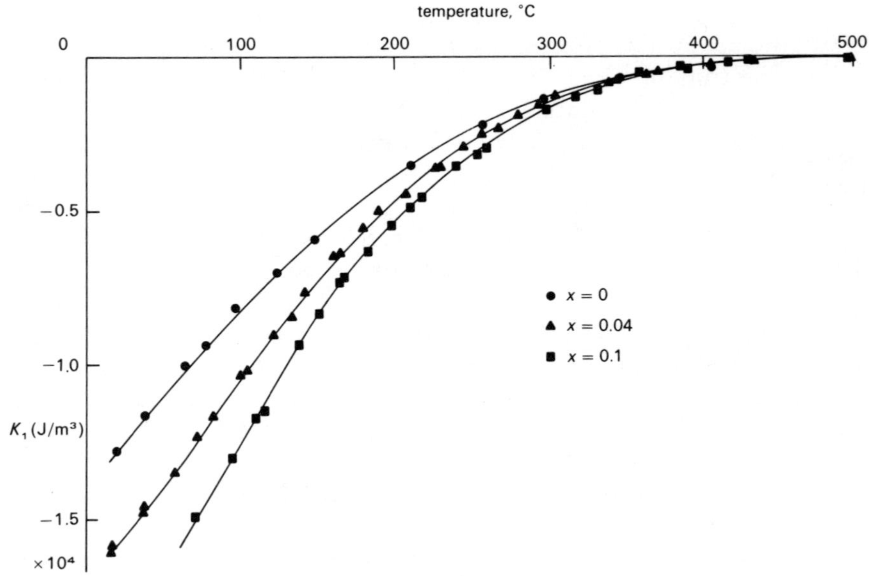

Figure 7.2 The variation of the cubic crystalline anisotropy constant K_1 for magnetite Fe_3O_4 between room temperature and the Curie point. The temperature dependence of $Fe_{2.96}Ti_{0.04}O_4$ and $Fe_{2.9}Ti_{0.1}O_4$ is similar and the anisotropy constant of all three compositions can also be expressed in this temperature range in the form $K_1(T) \propto M_s^p(T)$ where $p \sim 8$.

and $x = 0.68$, while changing sign, K_1 presumably falls to zero and will become zero again near $x = 0.8$.

The single-ion mechanism for the anisotropy of Fe^{2+} and Fe^{3+} ions in spinel oxides was discussed in section 3.4.2, and the contributions of single-ion anisotropies to the collective anisotropy of an exchange-coupled assemblage of ionic species in section 3.4.4. The K_1 of titanomagnetites comes from four contributions due to the two species of ion and the two symmetries. In magnetite, and possibly low x value compositions, the problem may be simplified by having no Fe^{2+} in tetrahedral (A) sites and therefore only Fe^{3+}_A, Fe^{3+}_B and Fe^{2+}_B contributions. With recourse to theories for the contributions of Fe^{3+} ions and other data, the contribution of Fe^{2+} in octahedral symmetry can be inferred. The inferred contribution per ion is negative and increases with increasing Fe^{2+} concentration in these low x titanomagnetites. The contribution falls steadily with increasing temperature between room temperature and the Curie point. Below room temperature the Fe^{2+} contribution is not so well-behaved. The possibility of the positive contributions of Fe^{3+}_A and negative contributions of Fe^{3+}_B cancelling to produce a zero point in K_1 was discussed in section 3.4.4. This does not seem to be the case in the titanomagnetites, the K_1 compensation point apparently resulting from a change in sign of the contribution of Fe^{2+}_B at low temperature. Compositions with higher x value have large positive K_1, at least at low temperature and this presumably indicates a large positive contribution from the Fe^{2+} ions, now present in higher concentrations. Whether this is due to Fe^{2+}_A or Fe^{2+}_B or the combined effect of both, is not certain.

The progressive increase in concentration of a highly anisotropic ion, like Fe^{2+} in titanomagnetite, does not necessarily produce a correspondingly large effect on the magnetization process by making the material 'harder', except possibly at absolute zero. The introduction of Fe^{2+} also weakens the exchange interaction and reduces T_c. At some finite temperature of interest, say room temperature, the ratio T/T_c therefore increases and

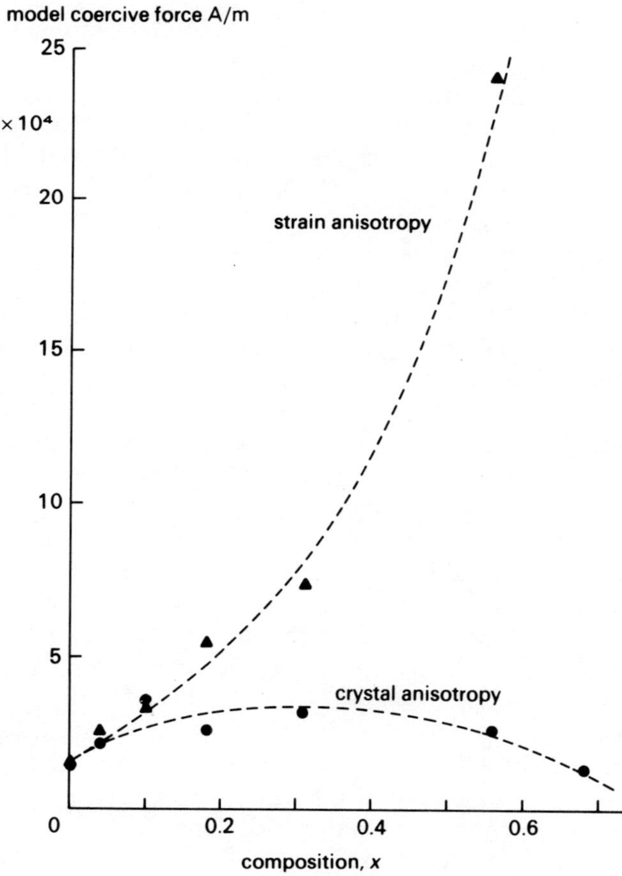

Figure 7.3 Anisotropy and magnetostriction in the titanomagnetites at room temperature. Model monodomain coercive forces are derived from the measured values of either K_1 or $\bar{\lambda}(=(2/5)\lambda_{100}+(4/5)\lambda_{111})$ at room temperature. A stress of 2×10^8 N/m^2 is assumed in the latter case. The dashed curves are not models but simply indicate the divergent trend in magnetic hardness due to the two sources. The principal cause of the rising hardness under the strain mechanism is the fall in M_s.

anisotropy and magnetization tend to have lower values than they would at a lower 'magnetic isotherm'. M_s is thus also affected, both directly and also indirectly by the change in magnetic isotherm. As Syono (1965) has pointed out, at room temperature the ratio $|K_1|/M_s$ has about the same value for $x = 0$ and $x = 0.6$ and the two compositions will have similar coercive forces in the monodomain state if magnetocrystalline anisotropy dominates. The susceptibility of $x = 0$ will, however, be higher because of the higher value of $M_s^2/|K_1|$. The hardest compositions lie near $x = 0.3$. Compositions with $x > 0.6$ would be the softest (see Fig. 7.3).

Magnetostriction and magnetocrystalline anisotropy are related (section 3.5), and if Fe^{2+} is a highly anisotropic ion when present in large concentrations, it is also a highly magnetostrictive one. The magnetostriction constants λ_{100} and λ_{111} have been measured for single crystals of titanomagnetite by Syono (1965) for a range of composition between 80 and 300°K. Again the effect of the introduction of the highly magnetostrictive ion is offset at room temperature by the lowering of the Curie point. Thus λ_{111} at room temperature lies in the range $+(80 - 100) \times 10^{-6}$ for compositions between $x = 0$ and $x = 0.68$, although at low temperature (80°K) it varies between $+12 \times 10^{-6}(x = 0)$ and $+330 \times 10^{-6}(x = 0.56)$. λ_{100} shows a greater variation, ranging from $-20 \times 10^{-6}(x = 0)$ to $+170 \times 10^{-6}(x = 0.56)$ at room temperature, and $-20 \times 10^{-6}(x = 0)$ to $+990 \times 10^{-6}(x = 0.56)$ at 80°K. An average magnetostriction, $\bar{\lambda}$, appropriate to polycrystalline material or an assemblage of crystallites, is provided by the weighted combination $(2/5)\lambda_{100} + (3/5)\lambda_{111}$. The room temperature $\bar{\lambda}$ for the measured compositions are: $x = 0$, $\bar{\lambda} = 2.9 \times 10^{-5}$; $x = 0.04$, $\bar{\lambda} = 5 \times 10^{-5}$; $x = 0.10$, $\bar{\lambda} = 5.9 \times 10^{-5}$; $x = 0.18$, $\bar{\lambda} = 8.4 \times 10^{-5}$; $x = 0.31$, $\bar{\lambda} = 8.9 \times 10^{-5}$; $x = 0.56$, $\bar{\lambda} = 1.39 \times 10^{-4}$. The model coercive force of an assemblage of monodomains with uniaxial strain anisotropy is $0.958(3/2) \times \bar{\lambda}\sigma/\mu_0 M_s$ (sections 3.5 and 4.3.1) and it can be seen that, at room temperature, the magnetic hardness of grains which have dominant strain anisotropy will increase markedly with increasing x because of the combination of the the increase in $\bar{\lambda}$ and (principally) the decrease in M_s. Figure 7.3 shows the variation of $3\bar{\lambda}/\mu_0 M_s$ with composition. σ, the stress residing in the grains, is taken to be about 2×10^8 N/m^2, a value well within the capacity of a ceramic material and one which provides a satisfactory match between the model coercive force and that observed in finely ground titanomagnetite (Fig. 7.5 and 7.9). Also plotted in the figure is $0.64|K_1|/\mu_0 M_s$, the model coercive force for monodomain grains with dominant cubic crystal anisotropy. Although similar hardness would obtain from either mechanism for magnetite, the difference at high x is large—a factor of 10 at $x = 0.6$ with the assumed stress.

No data for the variation of $\bar{\lambda}$ above room temperature is currently available for titanomagnetites. Klapel and Shive (1975) found a linear fall of λ_{111} for magnetite between room and Curie temperatures.

7.1.2 *The magnetization process in titanomagnetites*

The parameters of the magnetization process are χ, H_c, H_{cr} and M_{rs}. We may also add to this list W_{rp} and H_p, the peak rotational hysteresis loss per cycle and the field at which the peak loss occurs (Fig. 4.15). The parameters are functions of composition x (via material properties), grain size L (and also internal and surface microstructure), and temperature T (via temperature dependence of material properties, and thermal activation). While data covering the complete (x, L, T) field is not available, the (x, L, T_0) field has been covered, notably by Day *et al.* (1977) and Robins (1972). Other contributions to the overall picture of the magnetization process in the titanomagnetite system are provided by a number of workers using either synthetic materials or mineral samples. The compositions of the materials studied can be specified, provided they are single phase, to a required degree of accuracy in terms of a finite number of parameters—x and z in the case of a simple titanomagnetite. The extrinsic influences on the magnetization process—microstructure and mechanical strain—cannot be reduced to a simple parameter. The grain size quoted by experimenters is a statistic which represents the average dimension of a distribution of, presumably, more or less equidimensional grains. To speak of grain size, or some other geometrical parameter having relevance to the magnetization process, in the case of the highly irregular grains of magnetic minerals often found in rocks, is even more problematical.

The variability to be found in measured parameters of the magnetization process can, no doubt, largely be attributed to variations in undetermined extrinsic influences. A particular set of data can only be definitive if all variables are specified. No such data exists and the particular data of particular investigations presented here is to be regarded only as representative of the magnetization process in titanomagnetites.

7.1.2.1 *Composition dependence.* First we consider the compositional dependence of susceptibility at room temperature with grain size as a parameter. Representative data are presented in Fig. 7.4. The trend of this and other data is that susceptibility falls with increasing x, presumably due to the fall in M_s. This effect may be less pronounced in large grains.

The composition dependence of coercive force, M_{rs}/M_s and H_{cr}/H_c are shown in Figs. 7.5, 7.6 and 7.7 respectively.

The H_p values of the rotational hysteresis behaviour rise from about 7.6×10^4 A/m for magnetite to 2.4×10^5 A/m for $x = 0.6$ and fall for higher x (Manson *et al.*, 1979). Similar values are found in both fine and coarse grains. The peak loss W_{rp} is a more erratic function of composition and strongly grain-size dependent.

The overall trend in the data is an increasing hardness with increasing x, i.e. increasing Fe^{2+} concentration. Thus χ and H_{cr}/H_c fall whereas all

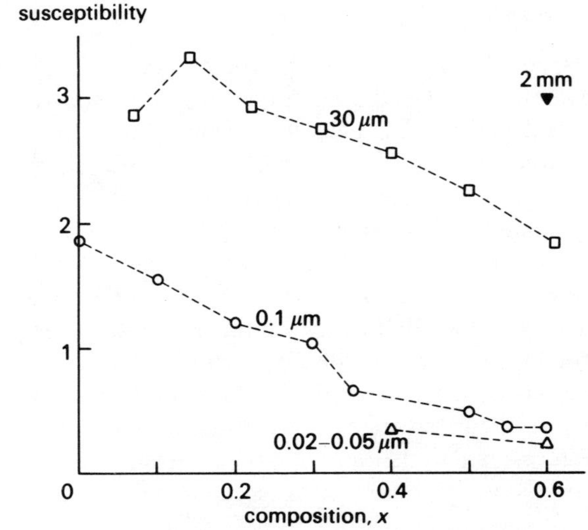

Figure 7.4 Representative data for the variation of room temperature volume susceptibility with composition x. The data of Day *et al.* (1977) for fine grain ball-milled material ($L \sim 0.1$ μm), of Robins (1972) for $L \sim 30 \pm 10$ μm, of Özdemir and O'Reilly (1981*a*) for $L \sim 0.02$–0.05 μm and Tucker and O'Reilly (1978) for a single crystal sphere ($L \sim 2$ mm) are included.

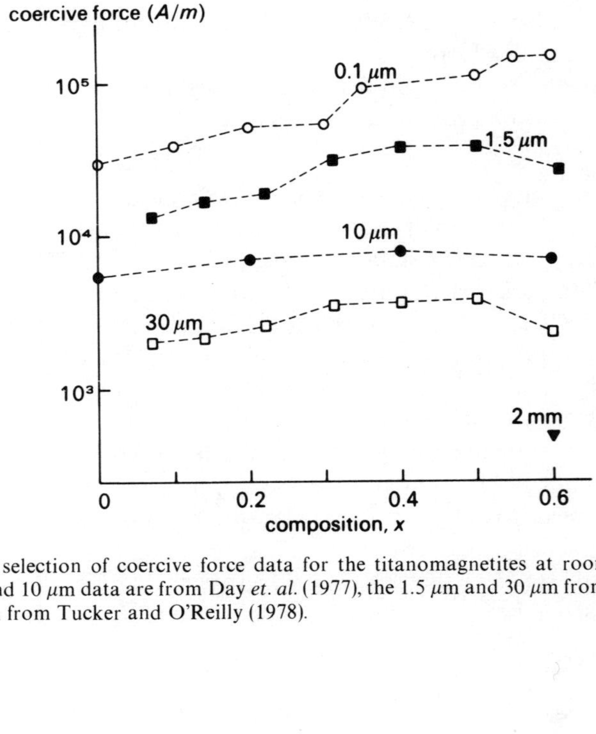

Figure 7.5 A selection of coercive force data for the titanomagnetites at room temperature. The 0.1 μm and 10 μm data are from Day *et. al.* (1977), the 1.5 μm and 30 μm from Robins (1972) and the 2 mm from Tucker and O'Reilly (1978).

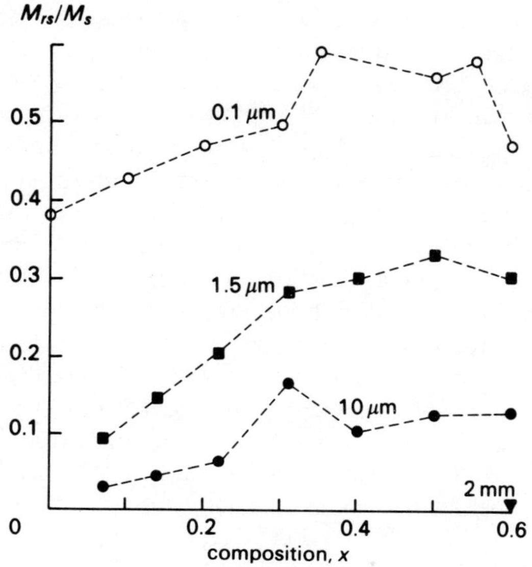

Figure 7.6 Variation of the room temperature saturation remanence to saturation magnetization ratio as a function of x for a number of particle sizes. Source of data: 0.1 μm (Day *et al.*, 1977), 1.5 μm and 10 μm (Robins, 1972) and 2 mm (Tucker and O'Reilly, 1978).

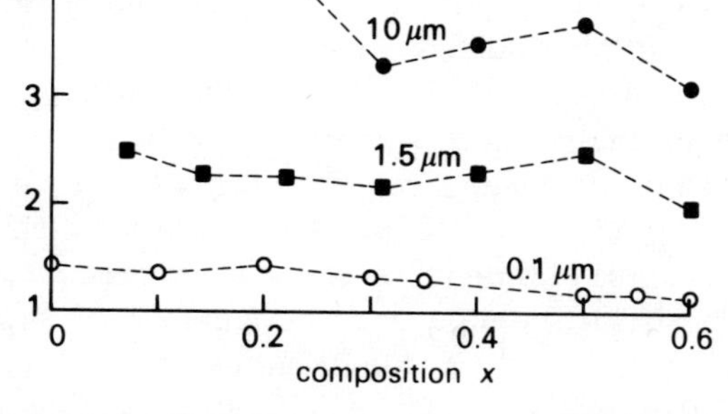

Figure 7.7 Variation of the room temperature coercive force of remanence to coercive force ratio as a function of x for a number of particle sizes. For the source of data, see caption to Fig. 7.6.

other magnetization process parameters rise. The data sets in which the increasing hardness is more pronounced presumably correspond to sample preparation procedures which result in a dominant strain anisotropy. This, combined with the fall in M_s, directly increases the coercive force of the model monodomain mechanism (Fig. 7.3). The internal strains can also bring about increased domain wall energy E_w and possibly increased E_w fluctuations. The first, again combined with the fall in M_s, will tend to raise the monodomain/multidomain limit and produce a more monodomain-like behaviour. The increased E_w fluctuation would also result in an increased hardness. Data sets in which the increase in hardness is less pronounced, or which show some softening towards $x = 0.6$ may have a larger relative contribution from cubic crystalline anisotropy. The high values of M_{rs}/M_s, in excess of the uniaxial anisotropy monodomain model value of 0.5, may be due to the inability of available fields to obtain a true saturation value M_s. M_s derived by extrapolation to infinite field can produce an M_{rs}/M_s ratio near to, or below, the model uniaxial value.

The substitution of Mg^{2+} for Fe^{2+} reduces magnetic hardness whereas the substitution of Fe^{3+} by Al^{3+} has more complicated results, at least at room temperature. This latter effect is because Al^{3+}, replacing Fe^{3+}, has a greater influence on Curie temperature and while the presence of Al^{3+} will generally increase the hardness at constant T/T_c it produces an increased hardness in $x = 0.4$ at room temperature and a decreased hardness in $x = 0.6$ (see section 7.1.2.3).

7.1.2.2 *Particle size dependence.* The particle size dependence of the magnetization process parameters is shown in figures 7.8 to 7.11. The sources of the data are Day *et al.* (1977), Dunlop (1973*a*), O'Donovan and O'Reilly (1982), Özdemir and O'Reilly (1981*a*), Parry (1965), Robins (1972) and Tucker and O'Reilly (1978). The general trend—that magnetic hardness increases with decreasing grain size—is clear. The parameters at the fine particle end of the size range are interpretable in terms of coherent rotation of monodomain grains with uniaxial anisotropy, the source of the anisotropy being strain in the case of compositions $x \gtrsim 0.4$. The dominant anisotropy in compositions with lower x could come from one or all of crystal, strain and shape sources. In the case of elongated particles, shape anisotropy may dominate although the parameters may have to be interpreted in terms of incoherent rotation of magnetization. Fine particle assemblages may contain a superparamagnetic fraction resulting in a softening of measured parameters. No sample of titanomagnetite has yet been prepared with small enough particle size to contain a preponderance of superparamagnetic material at room temperature. Such samples would presumably have particle sizes $\lesssim 0.01$ μm. More data is needed between 0.01 μm and 1 μm if the existence is to be established of a size range within which size-independent monodomain-like properties are found. At the upper end of the

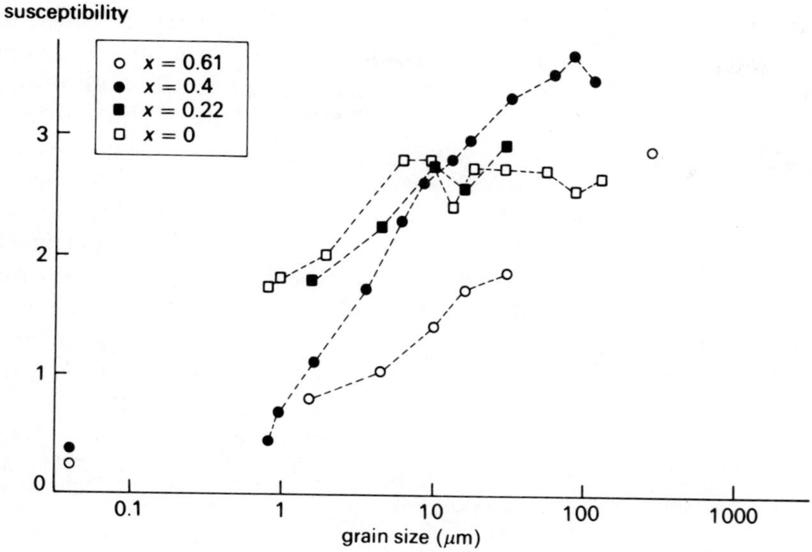

Figure 7.8 Variation of initial susceptibility with grain size at room temperature for titanomagnetites. Source of data: $x = 0$ and $x = 0.4$, Day *et al.* (1977); $x = 0.22$ and $x = 0.61$, Robins (1972); fine particle $x = 0.4$ and 0.6, Özdemir and O'Reilly (1981*a*); 2 mm single crystal sphere $x = 0.6$, Tucker and O'Reilly (1978).

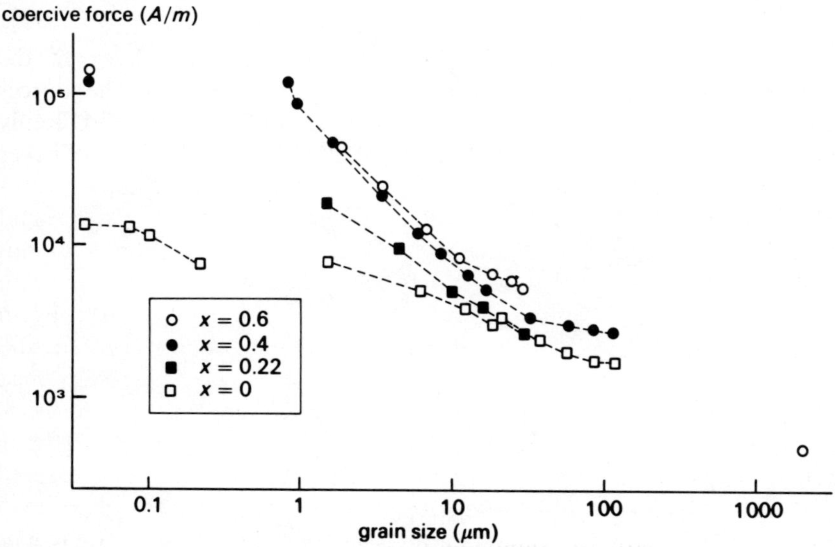

Figure 7.9 Variation of coercive force with grain size for titanomagnetites at room temperature. Source of data: fine particle (<0.25 μm) $x = 0$, Dunlop (1973*a*); coarser $x = 0$ (~ 1 to 100 μm), Parry (1965); $x = 0.4$, Day *et al.* (1977); $x = 0.22$, Robins (1972); $x = 0.6$, O'Donovan and O'Reilly (1982); fine particle $x = 0.4$ and 0.6, Özdemir and O'Reilly (1981*a*); 2 mm single crystal sphere $x = 0.6$, Tucker and O'Reilly (1978).

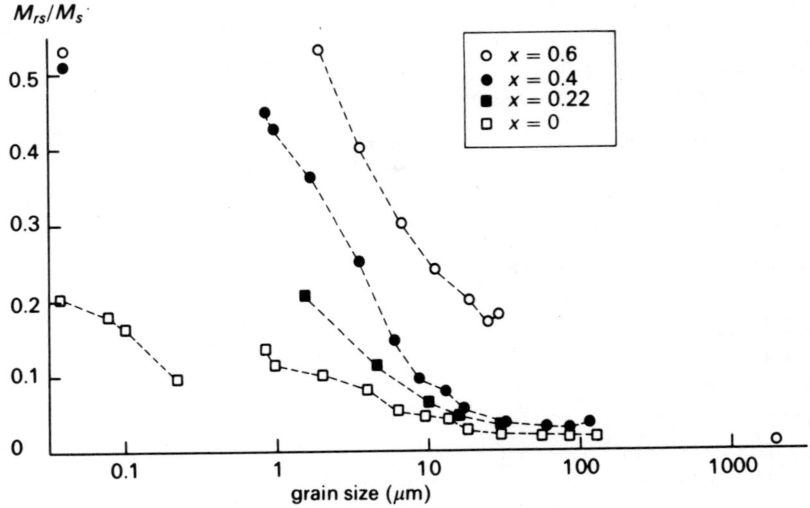

Figure 7.10 Ratio of room temperature saturation remanence to saturation magnetization as a function of grain size for titanomagnetites. Source of data: $x = 0$ (fine particle <0.25 μm), Dunlop (1973*a*); coarser $x = 0$ (~ 1 to 100 μm) and $x = 0.4$, Day et al (1977); $x = 0.6$, O'Donovan and O'Reilly (1982); fine particle $x = 0.4$ and 0.6 Özdemir and O'Reilly (1981*a*); 2 mm single crystal sphere $x = 0.6$, Tucker and O'Reilly (1978).

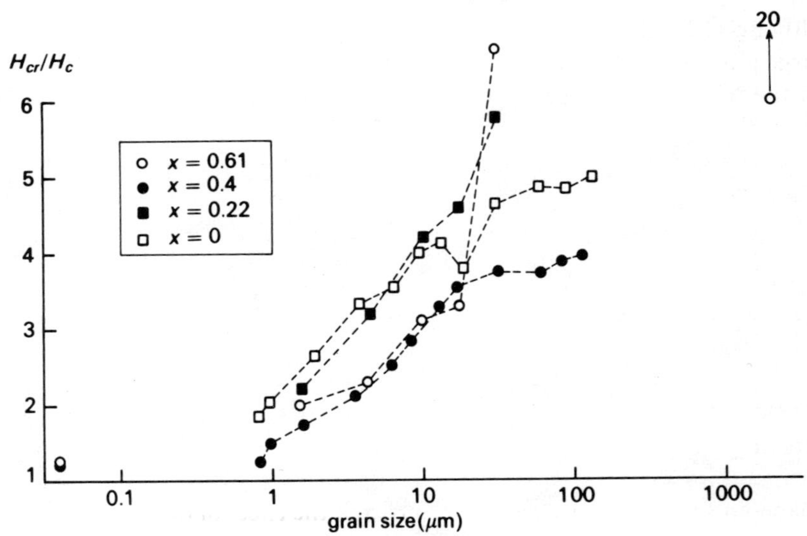

Figure 7.11 Variation of room temperature coercive force of remanence to coercive force ratio (H_{cr}/H_c) with grain size for titanomagnetities. Source of data: $x = 0$ and $x = 0.4$, Day *et al.* (1977); $x = 0.22$ and 0.61, Robins (1972); fine particle $x = 0.4$ and 0.6, Özdemir and O'Reilly (1981*a*); 2 mm single crystal sphere $x = 0.6$, Tucker and O'Reilly (1978).

size range the magnetization process can be described in terms of domains and domain walls and indeed the process of domain wall motion observed under the microscope. Observed coercive forces can be modelled in terms of voids or non-magnetic inclusions and regions of internal strains. The bulk of the data available lies in the range 1 to 100 μm and all properties vary more or less systematically in this range. If there is a distinct pseudosingle-domain regime forming a bridge between size-independent monodomain and size-independent multidomain behaviour it apparently embraces the whole of this 1 to 100 μm range.

7.1.2.3 *Temperature dependence.* A knowledge of the variation of magnetization process parameters above room temperature is of importance in understanding events occurring in nature. The variation below room temperature is of less direct relevance to the geophysical context but is important in laboratory experiments aimed at understanding the complex behaviour of minerals and rocks.

Above room temperature, the available hysteresis parameter data is found in the papers of Dunlop (1973*b*), Tucker and O'Reilly (1980) and Özdemir and O'Reilly (1981*a*). The dependence of the coercive force on temperature is often expressed in an indirect manner by a relationship of the form $H_c \propto M_s^n(T)$. A wide range of values of n have been observed and, for a given sample, different temperature ranges may be best fitted by different values of n. The variation of M_s with temperature can often be expressed satisfactorily by $M_s \propto (T_c - T)^m$ over a wide range of temperature, although the value of m may vary from sample to sample. This is the case for large grain single crystal titanomagnetite ($x = 0.6$) for which values of m lie in the range 0.4–0.5. This would suggest that, if $H_c \propto M_s^n(T)$ is valid then $H_c \propto (T_c - T)^q$ may also provide a useful description of experimental data. One complication is that fine particle assemblages may have an unblocking range extending well below the Curie point. As the unblocking range is entered the near-unblocked fraction reverses readily and the observed H_c is rather less than that of the stable fraction at that temperature.

In Dunlop's (1973*b*) suite of magnetite samples of mean grain size 0.22 μm, 0.10 μm, 0.076 μm, and 0.037 μm, the relationship $H_c \propto M_s^n(T)$ provides a satisfactory description of the data with n equal to 1.47, 1.79, 1.82 and 1.97 respectively. The last figure, for the smallest particle size, only holds for $T < 425$°C as, above this temperature, H_c falls more rapidly due to unblocking. A model for this data has been obtained in terms of both the temperature dependence of material properties (M_s in these presumed shape-anisotropy dominated materials) and the effect of thermal fluctuations on the coercive force (sections 4.4.4. and 6.1.3) (Dunlop and Bina, 1977). Thermal enhancement of magnetic susceptibility near the Curie point due to thermal fluctuations (the Hopkinson effect) as been widely reported in fine grain magnetite (e.g. Dunlop, 1974).

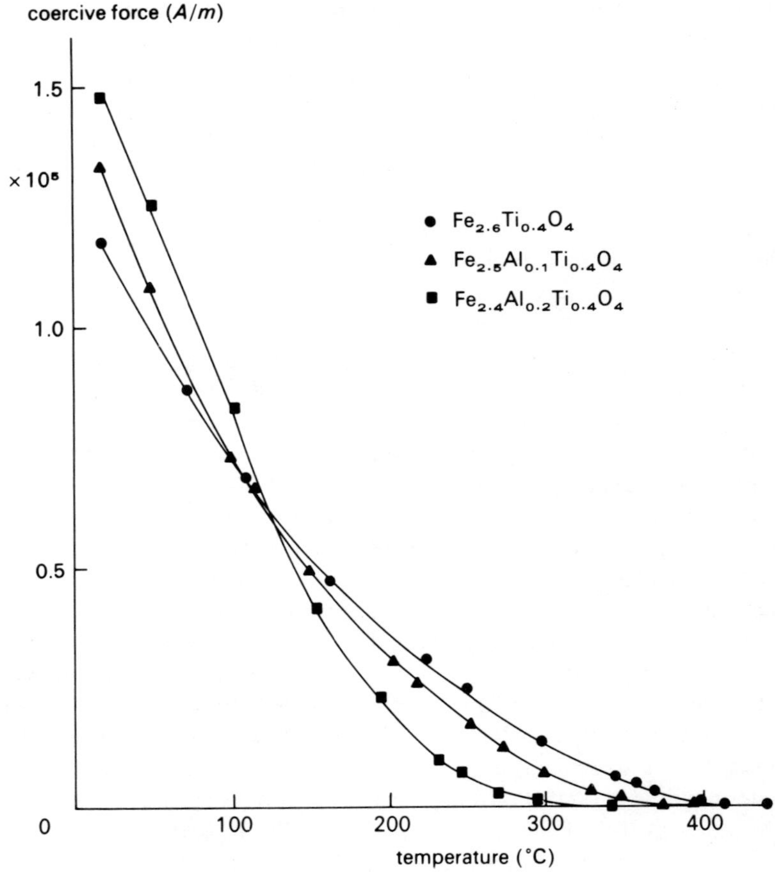

Figure 7.12 Variation of coercive force with temperature for fine particle (0.02–0.05 μm) samples of $Fe_{2.6}Ti_{0.4}O_4$, $Fe_{2.5}Al_{0.1}Ti_{0.4}O_4$ and $Fe_{2.4}Al_{0.2}Ti_{0.4}O_4$. This data can only be represented by a function of the form $H_c \propto M_s^n(T)$ when n takes different values in high and low temperature regions.

Data for fine particles (0.02–0.05 μm) $x = 0.4$ and $x = 0.6$ are not so well expressed by the $H_c \propto M_s^n(T)$ relation, the $\log H_c - \log M_s$ plot requiring two straight line sections. The tabulated data may be found in Özdemir and O'Reilly (1981*a*). The H_c data for the Al-substituted $x = 0.4$ series is shown in Figs. 7.12 and 7.13. This data is chosen to illustrate the influence of material properties on the temperature dependence of the magnetization process. The plot of H_c against temperature of Fig. 7.12 shows the complex composition/temperature dependence as the curves for the three compositions cross over. The complexity is removed on introducing the material parameter T_c by plotting H_c against magnetic isotherm T/T_c for the system (Fig. 7.13). The complex $H_c - T$ behaviour is then seen to arise from the effect of the compositional change on two material properties, T_c and

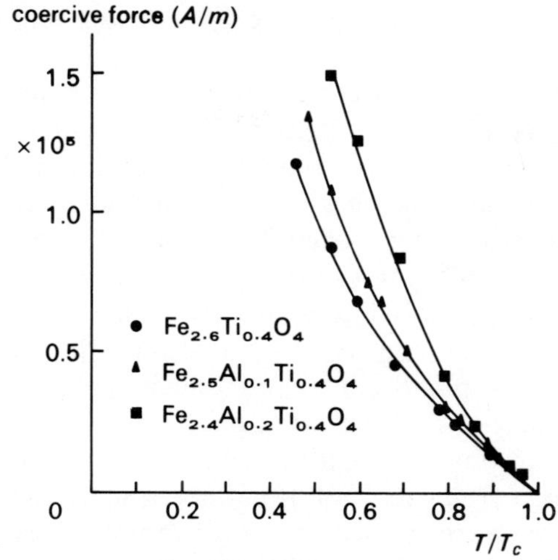

Figure 7.13 The influence of the material properties on the temperature dependence of the magnetization process is illustrated by replotting the data of Fig. 7.12 in terms of magnetic isotherms T/T_c.

(presumably), $\bar{\lambda}$, the first being lowered and the second raised by increasing Al^{3+} concentration.

The temperature dependence of the coercive force of large grain $x = 0.6$ is best fitted by an expression of the form $H_c = A(T_c - T)^{1-m} + B(T_c - T)^{7m}$, the constants A, B and m to be determined empirically for each specimen (Tucker and O'Reilly, 1980) and the two terms representing the influence of the temperature dependence of $\bar{\lambda}/M_s$ and K_1/M_s respectively. An example of the data is given in Fig. 7.14. The susceptibility of large grain titano-magnetite rises with increasing temperature, the susceptibility of $x = 0.65$

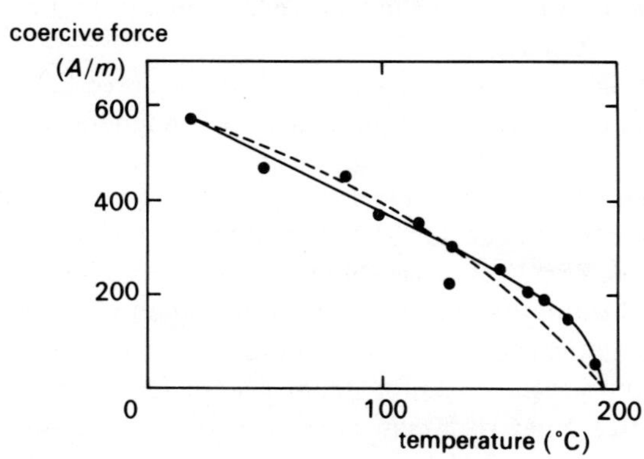

Figure 7.14 The temperature variation of coercive force in large grain (2 mm) $x = 0.60$. The dashed line is a power law of the form $H_c \propto (T_c - T)^q$ and the solid line $H_c = A(T_c - T)^{1-m} + B(T_c - T)^{7m}$, the two terms representing the influence of the temperature dependence of $\bar{\lambda}/M_s$ and K_1/M_s respectively.

F

reaching 180% of its room temperature value before falling to zero at the Curie point (Tucker, 1981).

The behaviour of titanomagnetites between room temperature and liquid nitrogen temperature is marked by a large increase in magnetic hardness with decreasing temperature in compositions and samples of which the properties are governed by magnetocrystalline anisotropy or magnetostriction. Thus the coercive force of large grain samples near $x = 0.6$ increases from $\sim 10^2$ A/m to $\sim 10^4$ A/m (Tucker 1981) and fine grain samples from $\sim 10^5$ A/m to values possibly in excess of 10^6 A/m (Özdemir and Banerjee 1981). Susceptibilities fall and M_{rs}/M_s ratios (in large grains) rise as temperature falls. The zero points in K_1 can produce some interesting and useful properties. In magnetite the zero point occurs at ~ 130°K and samples in which crystal anisotropy dominates the magnetization process, and which will usually be large grains or small near-spherical grains, shows the much-studied 'remanence transition' near this temperature and associated 'memory effect' (Nagata *et al.*, 1964). In a typical experiment a sample is magnetized to saturation at 80°K (or 300°K) and warmed (or cooled) in zero field. At the transition temperature a large fraction of the remanence may be lost depending on the size and shape of the grains. The transition temperature is, in effect, an unblocking temperature, the low anisotropy allowing the material to come into equilibrium with the zero ambient field or allowing the acquisition of a 'transition thermoremanence' if a field is present. If shape anisotropy dominates the magnetization of monodomain particles, the remanence transition is suppressed.

Being a material property, the zero point in K_1 and associated remanence transition is composition dependent and is depressed by the introduction of Ti^{4+} and variously affected by other cation species. The location of the remanence transition for compositions near $x = 0.6$ has been determined by Tucker (1981). For $x = 0.6$ the transition seems to be near 273°K, falling to 150°K for $x = 0.5$.

The observations of remanence transitions have useful but limited application in identifying the composition of magnetic minerals in rocks.

7.1.2.4 *Domain studies.* The observation of domain walls in titanomagnetites, the dependence of their spacing on particle dimension, their response to the application of an applied field, and preliminary observations of TRM-pinned domain walls, have been described in a series of papers by Soffel. A systematic study of the domain structure in synthetic titanomagnetite of compositions ranging from $x = 0.1$ to $x = 0.72$ and for particle sizes in the range 5 μm to 50 μm is described by Soffel and Appel (1982). The spacing, D, between domain walls falls as particle dimension, L, also falls. Plots of log D against log L are scattered (Fig. 7.15). This is presumably not solely due to experimental 'noise', but results partly from the fact that a single parameter L is only a rough measure of the influence of the dimensions

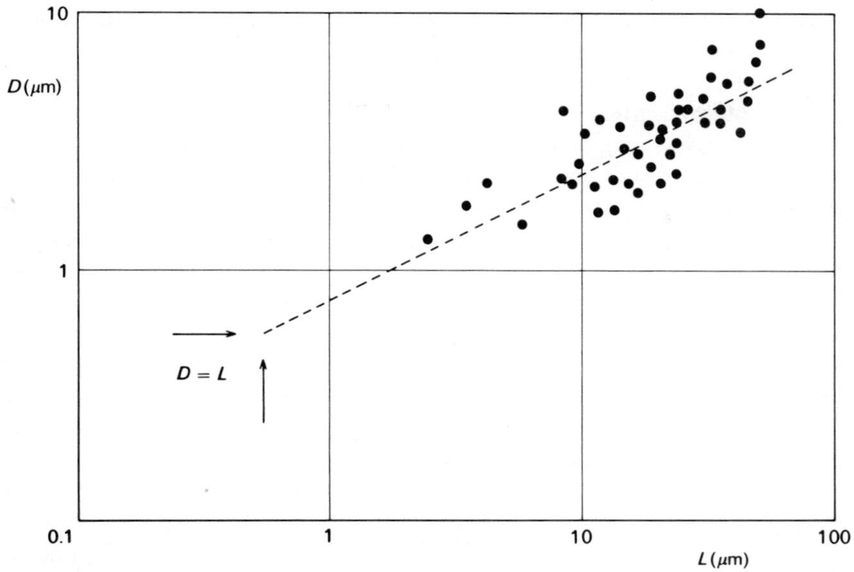

Figure 7.15 Domain structure observations in titanomagnetite grains of composition $x = 0.62$ (Soffel and Appel, 1982). D is the separation between domain walls and L is the particle size. The straight line represents an extrapolation to $D = L$, the size at which domain walls will no longer form.

of the real particle on domain structure, two dimensions only of a random section through the grain being observed. The internal microstructure must also perturb the domain wall spacing in a way which varies from grain to grain. We should therefore not expect a precise correlation between D and L. Extrapolating the $\log D - \log L$ plot to the point $D = L$ gives a measure of the particle dimension monodomain/multidomain boundary, which, for all compositions, is of the order of 0.5 μm. Again we should not expect a definitive, composition-dependent, value for this transition because of the difficulty in specifying the 'size' of a real grain unless it does happen, within experimental error, to be a cube or sphere, and because of the variation in other extrinsic or microstructural influences. For a given composition there may be a size above which the absence of domain walls is extremely unlikely and a smaller size below which walls do not form. The extent of the transitional size range may be indicated by the effect of the scatter of data points in Fig. 7.15 on the $D = L$ point. Within the transitional size range, some grains of a particular size may be monodomain while others of the same size are not. One might expect the proportion of the latter to the former to increase as the transitional size range is ascended. A further complication comes from the dependence of domain structure on magnetic history. Halgedahl and Fuller (1980) have observed the magnetization process in small titanomagnetite grains and found that in an assemblage of particles in the size range 5 μm to 30 μm a fraction does not nucleate domain walls after removal of a saturating field. The proportion of particles of a given size L which fails to nucleate a wall can be fitted to the function $A \exp(-BL^{\frac{1}{2}})$. A reverse field is required to nucleate a wall and when the nucleation field is sufficiently large the particle reverses by a single Barkhausen jump (see also section 4.3.3.1).

The general form of the domain structure in the titanomagnetite may be characterized by a preponderance of lamellar type domains separated by 180° walls. Low x (high M_s) values have the more complicated patterns with closure structure including pine-tree shaped closure domains. As x increases (and M_s falls) the closure structures become less of a feature of the patterns. For $x \sim 0.6$ the lamellar pattern is still predominant but the walls are less straight. Long spike domains extend well into the body of the grains.

7.1.2.5 *Modelling the magnetization process in titanomagnetites.* The material properties of section 7.1.1 can be combined with the model magnetization processes of Chapter 4 and the results compared to the measured properties reviewed in this section. To some extent, this has already been done in the course of our review of the experimental data and such comparisons are a feature of most reported experimental investigations. Because of the difficulty of directly determining what factors are influencing the magnetization process in a given material (even a well-characterized synthetic)—whether domain walls are present or not, whether shape, strain or crystal anisotropy dominates, to what extent thermal activation is operating—the modelling of the magnetization process has assumed some importance. The first task is to calculate a monodomain/multidomain transition size. Butler and Banerjee (1975) have calculated the room temperature monodomain/two-domain transition for cubic particles and rectangular parallelepipeds. The calculated transition size for cubes rises from 0.08 μm ($x = 0$) to 0.4 μm ($x = 0.7$). For parallelepipeds with elongation ratio 2.5 the corresponding particle lengths are 0.4 μm and 2 μm. These values seem to be in satisfactory agreement with the experimental ranges inferred from the Soffel and Appel (1982) data of the previous section. The ideal grains of the model calculation have no internal microstructure and qualitatively the inclusion of such microstructures could easily be imagined as raising or lowering the critical size in individual grains.

Model grains below the model critical size are subject to the rotation of magnetization mechanism and, for a small enough volume/temperature ratio, will be significantly affected by thermal agitation. The stable monodomain/superparamagnetic threshold depends on the chosen time constant and the origin of the anisotropy. For cubic grains, crystal anisotropy and a time constant of 1 s, Clark and Schmidt (1982) have obtained values ranging from 0.4 μm ($x = 0$) to 0.06 μm ($x = 0.7$). Calculations have also been carried out for 80°K where transition sizes vary less regularly because of the zero points in K_1 for $x = 0$ and $x = 0.56$. In the composition range between 0.1

and 0.3, the stable monodomain particle size range may be small or nonexistent at 80°K.

The temperature dependence of susceptibility has been modelled for equidimensional monodomain grains with crystal anisotropy, and elongated grains in which, at room temperature, shape and crystal anisotropy augment one another (Clark and Schmidt, 1982). Hopkinson peaks in susceptibilities appear at unblocking temperatures, the thermal enhancement of susceptibility being about 25 (section 4.4.4, Fig. 4.19(*a*)) when uniaxial (shape) anisotropy dominates and, when cubic crystal anisotropy is dominant, 190 ($K_1 < 0$) or 95 ($K_1 > 0$). For 0.06 μm equant magnetite grains, the peak susceptibility is about 1500 at an unblocking temperature of 130°C. A 10:1 elongated grain of length 0.06 μm peaks just below the Curie point. Near-equidimensional grains also show peaks near the zero point in K_1. The susceptibility of near spherical $x = 0.68$ of size 0.4 μm peaks at 3600 at about 70°C.

The effect of combined crystal, shape and strain anisotropy becomes rather complicated but can in principle be modelled under the assumption of relative orientations of the symmetry axes and magnitude of internal stress. The properties of fine grain titanomagnetite prepared by pulverization seem best matched by a model in which stress-derived uniaxial anisotropy dominates the magnetization process (Özdemir and O'Reilly, 1981*a*). The presence of such stress also significantly lowers the model superparamagnetic threshold to values of the order of 0.01 μm.

A model temperature variation of the susceptibility of 10 μm multidomain titanomagnetite grains has been determined by Clark and Schmidt (1982). For $x = 0$ to $x = 0.3$, the model susceptibility is approximately constant between about 250°C and the Curie point, tending to fall at lower temperatures. For $x = 0.056$ and $x = 0.68$, susceptibility falls more or less steadily as temperature decreases from the Curie point. The hysteresis properties of multidomain titanomagnetites modelled at low temperature (Tucker, 1981; Clark and Schmidt, 1982) are dominated by the large increases in λ and K_1 below room temperature. The high values of H_c ($\sim 4 \times 10^5$ A/m), M_r/M_s ($\sim$0.3–0.5) and values of H_{cr}/H_c approaching unity are not believed to be indicative of the monodomain state but are explicable by multidomain models with the appropriate material parameters.

Other examples of the modelling of the magnetization process parameters of magnetite and titanomagnetite may be found throughout the literature. No catastrophes seem to occur, presumably because it is always possible to choose values for indeterminate extrinsic factors that are both judicious and plausible.

7.1.3 *Weak field remanences in titanomagnetites*

The bulk of the reported work under this heading naturally involves TRM. ARM has been studied in magnetite and titanomagnetite, and magnetite

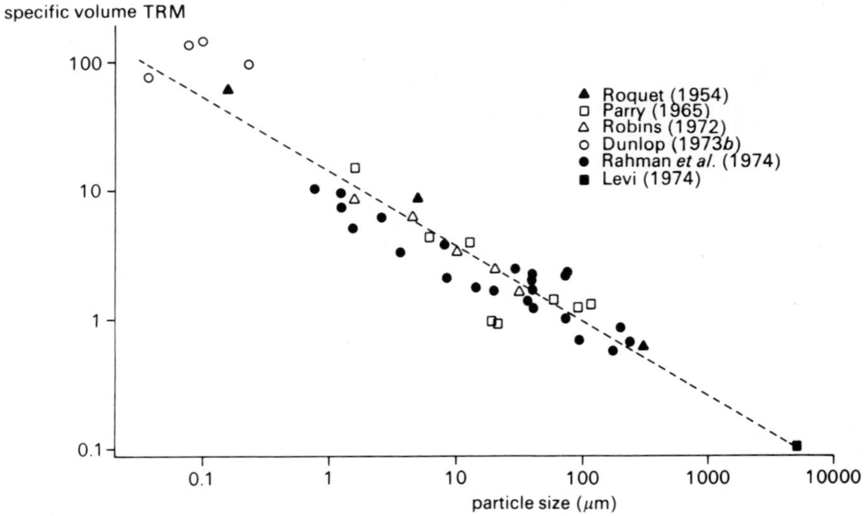

Figure 7.16 The slope of the low field, TRM (per unit volume)–H acquisition curves of magnetite as a function of grain size. A steady trend is observed throughout the entire size range. The dashed line has slope −0.6 (Dunlop, 1981).

has also figured as the magnetic carrier in several DRM investigations.

TRM acquisition curves are approximately linear in the weak field region (up to ~100 A/m, say). The slope of the TRM-field curve is dimensionless when the TRM is expressed as moment per unit volume. The values of the slopes of the weak field TRM acquisition curves for magnetite are shown in Fig 7.16 where the data from a number of investigations on a variety of grain sizes have been summarized by Day (1977) and Dunlop (1981). A steady rise in specific TRM with decreasing grain size is evident with no indication of separate monodomain, pseudo-single-domain or multidomain regions. The dashed line on the figure has a slope of −0.6, i.e. $\mathrm{TRM}/H \propto L^{-0.6}$ (Dunlop 1981). The specific weak field volume TRM for a selection of titanomagnetites is plotted in Fig. 7.17.

Expressions for the constants of proportionality for weak field TRM are provided by the models of section 5.2. Comparison of the observed and measured constants of proportionality require a knowledge of T_B, $M_s(T_B)$ and v when monodomain models are appropriate. Precise modelling requires that the particle volume distribution and distributions of T_B and $M_s(T_B)$ are taken into account. $M_s(T)$ is a strongly varying function near the Curie point and the uncertainty in $M_s(T_B)$ is correspondingly high when T_B approaches T_c. The unblocking temperature, determined by thermal demagnetization, is considered to be a close approximation to the weak field blocking temperature of the laboratory TRM. The multidomain TRM model also requires a knowledge of $M_s(T_B)$. The general and uniform trend of the particle size dependence suggested by Figs. 7.16 and 7.17 has yet to be modelled

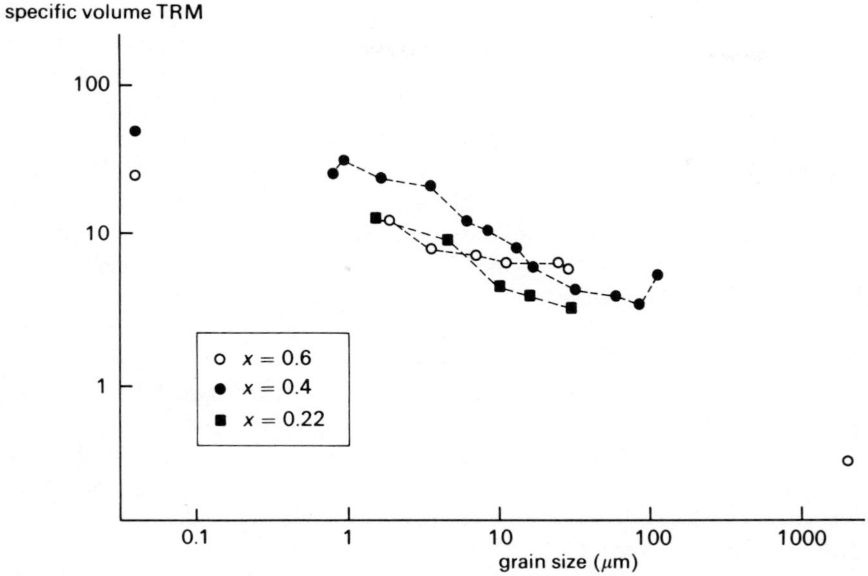

Figure 7.17 The slope of the weak field TRM–H acquisition curve for titanomagnetite of a number of compositions as a function of grain size. The $x = 0.22$ data is from Robins (1972), the $x = 0.4$ 1–100 μm data from Day (1977), the fine-particle $x = 0.4$ and $x = 0.6$ data from Özdemir and O'Reilly (1982*a*), the 1–30 μm $x = 0.6$ data from O'Donovan and O'Reilly (1982) and the 2 mm $x = 0.6$ data from Tucker and O'Reilly (1980).

satisfactorily. The monodomain model requires knowledge of $K_u(T)$ so that $T_B(K_u, v, \tau)$ can be evaluated. An explicit particle size dependence for the two-domain model (section 5.8.3) has to be worked out. The model TRM residing in monodomain-like moments in multidomain particles (section 5.2.4) falls with increasing particle volume as L^{-1}, or stepwise, depending on the source of the moment.

Stabilities of weak field TRM against AF demagnetization are expressed in terms of the median destructive field (MDF), the peak AF needed to remove 50% of the remanence. The power of the TRM mechanism as a recorder of a stable record of weak magnetic fields is evident in the high values of MDFs of fine particle magnetite and titanomagnetites (25,000–100,000 A/m). In large soft crystals the stability is much lower (MDFs of 1000–3000 A/m), intermediate particle sizes having intermediate stabilities.

TRMs in higher fields are characterized by a non-linear dependence on field and, eventually, saturation; by deviations from the law of additivity of PTRMs (Levi, 1979; Tucker and O'Reilly, 1980); and by shifts of the PTRM spectrum (i.e. shifts in the blocking temperatures) to lower temperatures as the inducing field is increased (Clauter and Schmidt, 1981). In small grains the MDF of TRM falls as inducing field increases, i.e. the high field TRM is softer, whereas the opposite result holds for large grains (the Lowrie–Fuller

(1971) test). The model TRM-H acquisition curves for monodomain grains are shown in Fig. 5.1. Curve (b) (the random array of uniaxial monodomain grains) provides a reasonable fit for fine particle data for plausible values of the parameter a. The model of curve (b) has a defect in that the approximation that the grain moments lie only along the unique grain axes will become less satisfactory as applied field increases. An even better fit to fine particle data is provided by a model in which this approximation is not made, and the deviations of the moments from the easy axes calculated numerically (Schutts and Schmidt, quoted by Dunlop (1981)). Again more precise modelling will accommodate a range of particle volume and blocking temperature.

Large grains can exhibit linear segments in a bilogarithmic TRM-H plot, such as is predicted by the two-domain model of Fig. 5.3. In the case of titanomagnetite ($x = 0.6$) good agreement can be obtained between the observed 'break points' in the acquisition curve and a model which contains a distribution of microscopic coercive forces derived from AF demagnetization of PTRMs, two modes of domain wall pinning and a fraction of the remanence blocked isothermally (Tucker and O'Reilly, 1980). The Königsberger ratio for TRM, Q_T, the ratio of the specific weak field TRM to the susceptibility, rises steadily as grain size falls. Coarse grain magnetite and titanomagnetite have values ~ 0.1 and fine grain material ~ 100. The effect of cooling rate on intensity of acquired TRM—slow cooling enhancing fine-particle, and diminishing large-particle TRM—has not yet been systematically studied, but the existence of the effect has been demonstrated in (possibly magnetite-bearing) artefacts (Fox and Aitken, 1980).

ARM in magnetite has been studied by a number of workers. Levi and Merrill (1976) determined ARM/TRM ratios for material with grain size in the range $\sim 0.1\ \mu$m to $3\ \mu$m and two crystals ~ 1 cm in size. In the first group of samples the ARM/TRM ratio lay between about 0.1 and 0.6, the large crystals had ratios of 1.5 and 2. Suguira (1979) found a definite concentration dependence of ARM in dilute dispersions of magnetite ranging from $2.6 \times 10^{-4}\,\%$ to 2.3%. TRM was relatively constant but the ARM/TRM ratio fell from about 0.8 in the most dilute sample to 0.1 in the most concentrated. In the 10%–30% concentration range ARM (and M_{rs}/M_s) also depend on the overall shape of samples containing 0.2–0.4 μm magnetite particles (Schmidbauer and Veitch, 1980). Although a concentration dependence of ARM is observed in such samples, the law of additivity of PARMs seems to hold, as does the equality of ARM lost, during AF demagnetization, to ARM gained in the same AF interval (analogous to the equivalence of thermal demagnetization of TRM and PTRM acquisition of section 5.3). The normalized weak field ARM is more stable to AF demagnetization than the saturation remanence but has similar stability to TRM. Facey (1983) has compared ARM and TRM intensities in 50 μm titanomagnetites of a number of compositions. Within experimental error, the TRM/ARM ratio was found to be equal to the ratio $M_s(T_0)/M_s(T_B)$. The AF demagnetization curves crossed over, TRM being more stable than ARM in fields less than about 16 000 A/m.

A catalogue of DRM studies, in some of which magnetite is the remanence carrier, is provided by Barton and others (1980). The material or other magnetic properties of the magnetite are probably less important than the variations in deposition procedure and the method of consolidating the sediment. In general the process seems effective in recording the direction of a weak ambient field.

7.2 The titanomaghemites

Simple titanomaghemites have a formula unit $Fe_{(3-x)R}Ti_{xR}\square_{3(1-R)}O_4$ where $0.1 < z < 1$, $(8/(8.1 + 0.1x) > R > 8(9 + x))$ and $0 < x < 1.285$ (section 2.1.2). Titanomaghemites with $x > 1$ (Fig. 2.4) have yet to be synthesized or found in nature. Titanomaghemites are metastable (Fig. 2.11) and have been prepared in the laboratory by the oxidation of titanomagnetites at temperatures below about 350°C. Oxidation is usually carried out in air and, if the oxidation product is single phase, the addition-of-oxygen mechanism (section 2.1.2) must apply. The problems of laboratory maghemitization involve preparation of suitable starting material (fine grain near stoichiometric single phase titanomagnetite), the determination of z, the degree of maghemitization, by some chemical analytical technique which is independent of any presumed variation of material properties with composition, and an adequately sensitive means (usually X-ray analysis) of detection of other phases in the oxidation product. In oxidation studies carried out in other media than air, e.g. water or hydrogen (oxidation in the latter medium clearly comes about through some other agent present in the system!), whether the addition-of-oxygen mechanism or removal-of-iron mechanism dominates has yet to be decided. The removed iron, which presumably rapidly becomes oxidized itself by addition of oxygen, may remain in intimate association with the oxidized titanomagnetite. Care may be needed to distinguish such a two-phase titanomaghemite-containing product from the multiphase product of some other oxidation process (Fig. 2.1). Representative accounts of how researchers in different laboratories have approached the problems of titanomaghemite synthesis may be found in Keefer and Shive (1981), Nishitani and Kono (1983) and Özdemir and O'Reilly (1981*b*).

7.2.1 *Intrinsic properties*

The intrinsic properties depend on composition and cation distribution and are, ideally, independent of microstructure. Taking again the assumptions of the cation distribution of section 7.1.1.1, i.e. Ti^{4+} and vacancies only in octahedral sites, an A–B collinear spin structure and spin-only moments, the saturation magnetization n_β is a function of x, z and δ, where δ describes the $Fe^{2+} - Fe^{2+}$ distribution under the above assumptions, and can be chosen to fit the experimental data. The model saturation magnetization may be related to the maghemitization process itself (O'Reilly, 1983) by assuming that in the surface oxidation/internal diffusion process there is a differential availability, due to different site binding energies, of Fe^{2+} in octahedral and tetrahedral sites. A parameter α describes the availability of tetrahedral site Fe^{2+} relative to octahedral site Fe^{2+}. δ, the number of Fe^{2+} per formula unit, becomes $\delta = \delta_0(1 - z)R/(1 - z + \alpha z)$ for the addition of oxygen mechanism and $\delta = \delta_0(1 - z')/(1 - z' + \alpha z')$ for removal of iron, where $\delta_0(x)$ describes the $Fe^{2+} - Fe^{3+}$ distribution in the starting titanomagnetite and z' is an alternative oxidation parameter (section 2.1.2 and Fig. 2.5).

The available low temperature saturation magnetization data is for material to which the addition of oxygen mechanism applies and a constant value of α (~ 0.2) seems to hold at least for the early stages of oxidation in which the magnetization falls with increasing z (e.g. Readman and O'Reilly, 1972). One problem is that the simple model 'intrinsic' properties correspond to an infinite crystal lattice. In the fine particle titanomaghemites ($\lesssim 0.1\ \mu m$) such a model may be inappropriate because of weakly exchange-coupled magnetic ions near the particle surface, and the spontaneous magnetization may be reduced. Some room temperature saturation magnetization data for

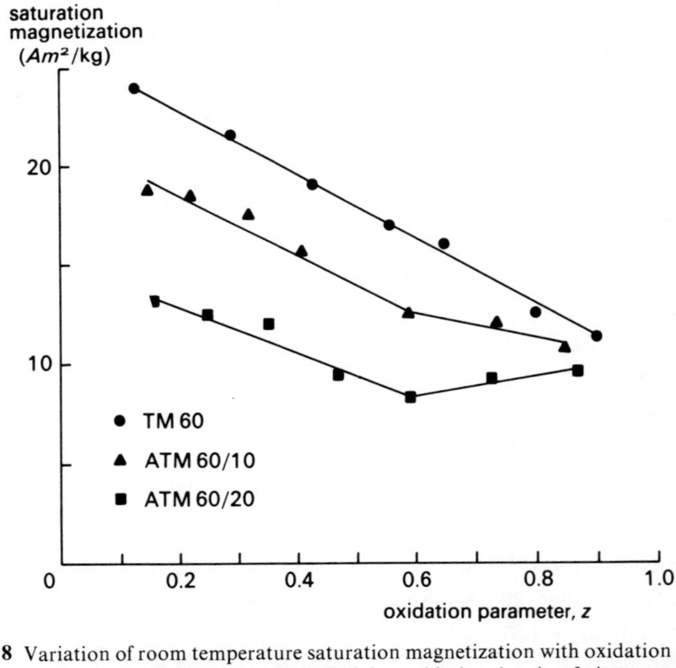

Figure 7.18 Variation of room temperature saturation magnetization with oxidation parameter z of three series of titanomaghemites produced by oxidation in air of titanomagnetites of composition $Fe_{2.4}Ti_{0.6}O_4$ (TM60), $Fe_{2.3}Ti_{0.6}Al_{0.1}O_4$ (ATM60/10) and $Fe_{2.2}Ti_{0.6}Al_{0.2}O_4$ (ATM60/20).

titanomaghemites of grain size 0.02–0.05 μm produced by oxidation of $Fe_{2.4}Ti_{0.6}O_4$, $Fe_{2.3}Ti_{0.6}Al_{0.1}O_4$ and $Fe_{2.2}Ti_{0.6}Al_{0.2}O_4$ (Özdemir and O'Reilly, 1982*b*) are shown in Fig. 7.18.

The Curie temperatures of titanomaghemites are of particular interest as providing a means by which, in conjunction with other data, the degree of maghemitization of titanomagnetites in rocks may be determined. Several investigations have produced Curie point data for the titanomaghemite field, and it has become traditional to express the results as contours of constant Curie temperature in the titanomaghemite quadrilateral of the ternary diagram. The fact that the contours from the different sources are not in close agreement has been the subject of some discussion. It should be remembered that the contours are a subjective interpolation of a finite number of data points; that there is uncertainty in both the composition coordinate and (more important) in the measured Curie temperature, the latter also depending on the convention adopted; and that the determination of Curie temperatures above about 350°C (the inversion temperature) are obtained by extrapolation and are associated with larger uncertainties. It may be that there is a 'true' set of contours and that the experimentally determined sets are scattered about this true set by experimental noise, both in material synthesis and measurement. There may also, however, be systematic differences. Each contour set is based on a suite of unoxidized titanomagnetites and the contours tied to the Curie points of the unoxidized

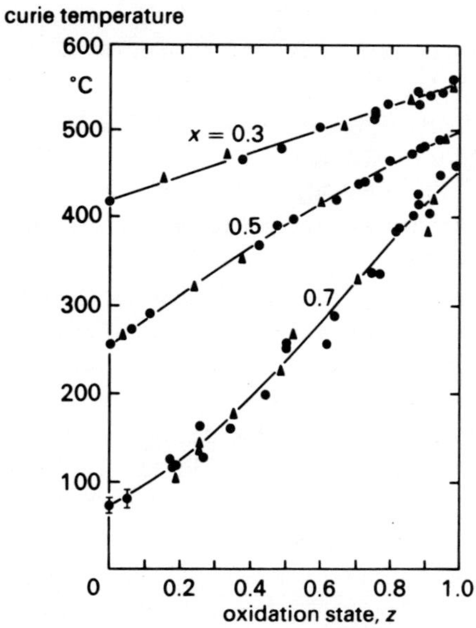

Figure 7.19 Variation of Curie temperature with oxidation parameter z of three series of titanomaghemites produced by oxidation in air of titanomagnetites with composition $x = 0.3$, 0.5 and 0.7.

suite. As noted in section 7.1.1.2, the Curie temperature of '$Fe_{2.4}Ti_{0.6}O_4$' can range between 150°C and 220°C, and T_c-z sets based on starting material at either and of this range will no doubt differ significantly. Rather than reproduce a set of contours here, it is more instructive to look at an original data set from which contours have been produced. Figure 7.19 shows the Curie temperatures of titanomaghemites prepared by Nishitani and Kono (1983) starting from compositions $x = 0.3$, 0.5 and 0.7. The open circles are samples for which the composition parameter z was determined directly, the remainder having their z coordinate inferred from a presumed cell edge-z relationship. All investigators agree that the Curie point temperatures rise as oxidation proceeds, the exchange interaction being enhanced by the increased Fe^{3+} concentration.

Direct determination of magnetostriction or magnetocrystalline anisotropy constants is not possible with the fine particle material available. Their variation with degree of maghemitization can only be inferred from the measured magnetization process parameters of the titanomaghemites.

7.2.2 *The magnetization process*

Titanomaghemites have been successfully prepared only in the fine grain form ($<1\ \mu$m) and the magnetization process parameters are therefore known only for such material. Like the equivalent fine grain titanomagnetite, coercive forces are high at room temperature except for highly maghemitized compositions. Figure 7.20 shows the room temperature coercive force and susceptibility for a suite of titanomaghemites of grain size 0.02–0.05 μm based on compositions $Fe_{2.4}Ti_{0.6}O_4$, $Fe_{2.3}Ti_{0.6}Al_{0.1}O_4$ and $Fe_{2.2}Ti_{0.6}Al_{0.2}O_4$. At room temperature the coercive force rises before eventually falling as maghemitization degree increases. Just as in the case of fine grain titanomagnetites (Fig. 7.12) the compositional dependence of coercive force is complicated by the variation of more than one material property, both T_c and the hardness parameter $\bar{\lambda}/M_s$ being functions of z. The initial hardening results from the increase in T_c with increasing z and the later softening from the decrease in the concentration of highly magnetostrictive or anisotropic Fe^{2+}. Measurements over a range of temperature allows H_c to be plotted against z at a constant magnetic isotherm, T/T_c, and such plots show a monotonic fall of H_c with increasing z (Özdemir and O'Reilly, 1982*b*). M_{rs}/M_s ratios for the material of Fig. 7.20 lie in the range 0.5 to 0.6 increasing as z increases. H_{cr}/H_c ratios lie between 1.1 and 1.3. The properties of these titanomaghemites are consistent with the monodomain state with uniaxial Fe^{2+}-related anisotropy.

The growth and decay of VRM has been observed in these fine grain titanomaghemites (Özdemir and Banerjee, 1981). When plotted in the form M_{VRM}–$\log t$, the acquisition and decay curves are approximately linear with slopes (viscosity coefficients) of 0.5 A/m for $z < 0.6$ rising to about 3 A/m

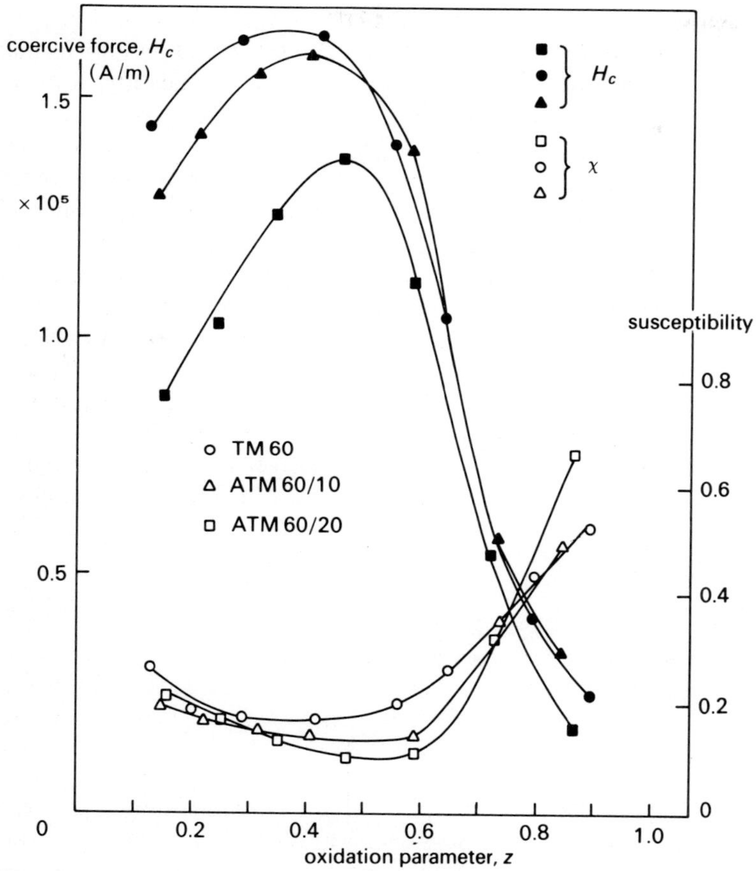

Figure 7.20 Variation of room temperature coercive force and susceptibility with degree of maghemitization of the titanomaghemite series of Fig. 7.18.

for the softest composition ($z = 0.9$). The particle size at the superparamagnetic/stable monodomain boundary at room temperature may be calculated from the data of figures 7.18 and 7.20. For $\tau \sim 10^2$ s this lies at $\sim 0.015\ \mu m$ for $0 < z < 0.6$, rising thereafter to about $0.03\ \mu m$ at $z = 1$. It may be that the boundary in material with a smaller contribution from magnetostriction-based anisotropy would lie at larger particle dimensions. The monodomain/multidomain boundary region may have a similar size range to titanomagnetites.

7.2.3 *The inversion of titanomaghemites*

As mentioned in section 2.1.2, the cation-deficient titanomaghemites invert on heating and, presumably (like maghemite), on application of hydrostatic

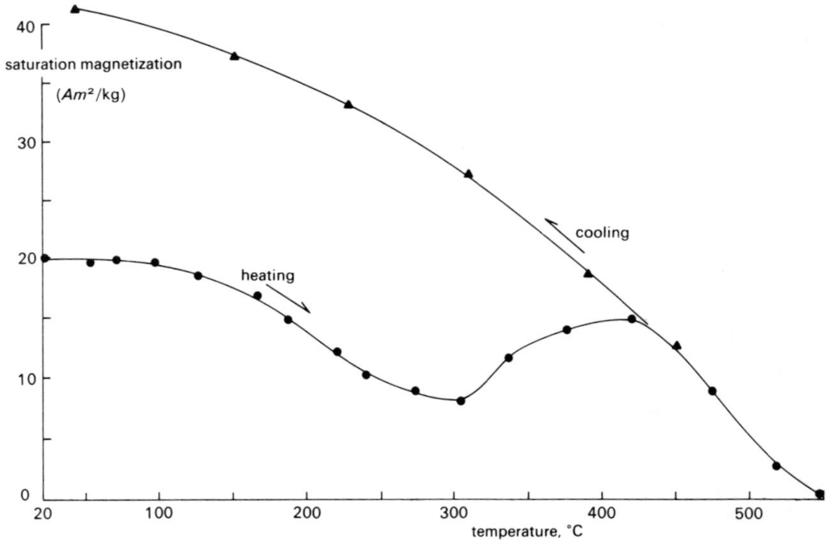

Figure 7.21 Temperature dependence of high field magnetization (the thermomagnetic curve) of a titanomaghemite ($x = 0.6$, $z = 0.29$) showing irreversibility due to inversion of the titanomaghemite during the heating part of the run. The magnetization after the run is about twice as large as that before.

pressure. This results in characteristic irreversible thermomagnetic (M_s–T) curves obtained when titanomaghemites are heated in an inert atmosphere, and such curves have proved useful in establishing the widespread occurrence of maghemitization in the pillow lavas of the ocean floors. An example of an irreversible curve for synthetic titanomaghemite ($x = 0.6$, $z = 0.29$) is shown in Fig. 7.12. The irreversible change often occurs at temperatures of about 350°C although higher inversion temperatures are also reported. The magnetization of the material after heating to elevated temperatures (M_F) is often higher than the initial magnetization before heating (M_I). Model values for the M_F/M_I ratio as a function of z can be calculated from the inversion model of section 2.1.2 (O'Reilly, 1983) and some measure of agreement is obtained with experimental data on synthetics (e.g. that of Nishitani and Kono (1983)). The measurement of M_F/M_I as a quantitative means of determining the degrees of maghemitization of titanomaghemites in rocks requires further development.

Maghemite is reported as inverting at 0°C under a pressure of 1.5×10^7 Pa (Kushiro, 1960) or at lower pressures at elevated temperature. No data is available for titanomaghemites, but the recovery of uninverted titanomaghemite-bearing rocks from depths below the ocean surface considerably in excess of 1500 m would seem to indicate that the stability to hydrostatic pressure is well in excess of the published maghemite figure.

7.2.4 *Weak-field remanences*

The growth of VRM in titanomaghemites has been mentioned in section 7.2.2. The acquisition of TRM in the same materials has been studied by Özdemir and O'Reilly (1983). Samples with blocking temperatures higher than the inversion temperature cannot acquire TRM. The specific weak field TRM intensity for a titanomaghemite suite based on a starting composition $Fe_{2.2}Ti_{0.6}Al_{0.2}O_4$ is shown in Fig. 7.22. The intensity falls approximately hyperbolically with degree of maghemitization, highly maghemitized compositions having a TRM intensity 10–20% of that in the unoxidized titanomagnetite. As in the case of the equivalent fine grain titanomagnetites (section 7.1.3) the TRM is highly stable with MDFs of the order of 10^5 A/m. The stability of the TRM of the maghemitized material over the composition range capable of acquiring TRM is greater than the equivalent titanomagnetite because of the rise in magnetic hardness over this range (Fig. 7.20).

It may be unlikely that titanomaghemites in nature acquire TRM, although complex substituted titanomagnetites with Curie temperatures or blocking

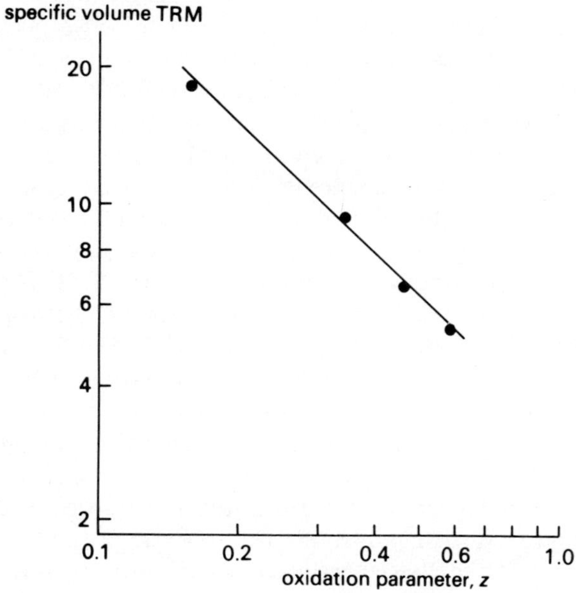

Figure 7.22 Specific weak field TRM intensity as a function of composition parameter z, for a suite of titanomaghemites of grain size $0.02-0.5$ μm. The relationship is approximately hyperbolic.

temperatures below room temperature may acquire a CRM, while maghemitization is bringing about an increase in T_c, and indirectly in anisotropy, on passing through the blocking composition z_B at the alteration temperature (section 6.1.1). What may be common in nature is that titanomagnetite acquires TRM during rapid cooling and this is later modified by maghemitization. According to the data from synthetics (Fig. 7.20) this first produces a hardening of the material so that the remanence becomes more stable. The statistical distribution of monodomain grain moments remains frozen in. The intensity, however, will fall because M_s is falling (Fig. 7.18), eventually reaching (say) 50% of the initial value. The remanence will soften in the later stages of maghemitization with the possibility of growth of VRM or total unblocking in small enough grains. Changes in microstructure due to splitting or cracking of grains (section 2.2) are also likely to modify the intensity and stability of remanence.

7.3 Multiphase products of the oxidation of titanomagnetite

Deuteric (high temperature) oxidation of titanomagnetites (section 2.2) beyond the region of high temperature non-stoichiometry results in a multiphase intergrowth. In general, the magnetic constituents of the inter growth differ both in composition and microstructure from the original titanomagnetite grain (except in the special case of martitization of magnetite, when the magnetic constituent of the intergrowth is also magnetite except when oxidation is complete). The subdivision of the original homogeneous titanomagnetite grain means that the magnetic constituents of the intergrowth have smaller grain dimension than the original grain, and this led to the suggestion that the divided grain would be associated with enhanced hardness and stability of remanence (Graham, 1953). Experimental evidence comes from the correlation between stability of NRM and effective grain size, including small grains formed by subdivision of larger grains, found for a wide range of rocks by Larson and others (1969). The progress of deuteric oxidation in titanomagnetites in rocks has been classified into stages based on the evolution of the characteristic textures and co-existing phases as observed under the optical microscope (Haggerty, 1976). The 'exsolution' stages are characterized by the progressive formation of ilmenite lamellae separating progressively titanium depleted titanomagnetite. In the later 'post-exsolution' stages, pseudomorphic replacement of the magnetite and ilmenite by higher oxides—haematite, pseudobrookite, and rutile—takes oxidation towards completion. The intrinsic properties of the intergrowth depend on the concentration and composition of the constituent oxides. The magnetization process may be complicated by interactions between the closely spaced residual magnetic regions.

7.3.1 *Intrinsic properties*

Deuteric oxidation has been simulated in the laboratory by heating large grains of titanomagnetite, $Fe_{2.4}Ti_{0.6}O_4$, at 1275°C under controlled atmosphere (Tucker and O'Reilly, 1980*b*). This is higher than the temperatures at which deuteric oxidation probably takes place in nature (600–1150°C) but intergrowth structures similar to those observed in titanomagnetites in rocks are obtained. Below $z \sim 0.1$ the grain consists of a single phase non-stoichiometric titanomagnetite and the Curie temperature rises steadily (Fig. 7.23). Beyond this region of high temperature non-stoichiometry, the exsolution stage ends at $z \sim 0.6$ (the oxidation parameter here describing the bulk composition of the intergrowth). The saturation magnetization, its temperature dependence and the observed Curie points can be ascribed to the presence of three phases in the intergrowth—two titanomagnetites and one haematite-ilmenite solid solution member. The magnetic properties are dominated by the two titanomagnetite phases. The concentration and compositions of the phases depend on the degree of oxidation. Initially the lower Curie point titanomagnetite, with x gradually falling from the starting value (~ 0.6) as oxidation proceeds, is the more abundant. The second

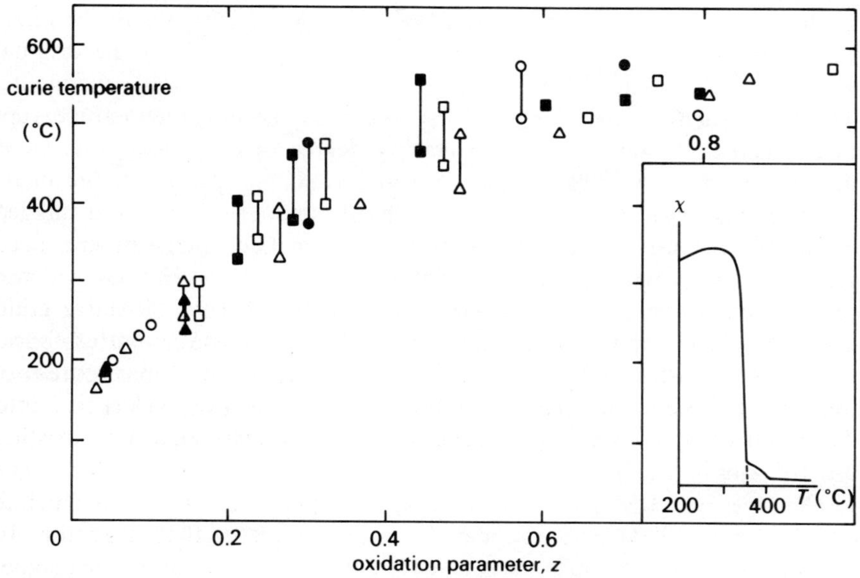

Figure 7.23 Variation of Curie temperature with bulk composition of a suite of deuterically oxidized titanomagnetite grains, ($x \sim 0.6$). A single Curie point (non-stoichiometric titanomagnetite) is observed in the region of high temperature non-stoichiometry, two Curie points (two titanomagnetites) in the exsolution stage of oxidation, and one Curie point (magnetite) in the post-exsolution, pseudomorphic replacement, stage. The inset shows the $\chi - T$ curve for one of the samples.

titanomagnetite (starting value $x \sim 0.8$) increases in concentration (and decreases in x value) until at $z \sim 0.4$ the two are present in roughly equal concentration. Both phases then fall to low concentration and to composition $x \sim 0$ at the end of the exsolution stage. In the post-exsolution stage of oxidation, magnetite, present at levels of a few percent and possibly the result of pseudomorphic replacement of the intergrown phases rather than lamellar subdivision, is the only magnetic constituent.

7.3.2 *The magnetization process*

Deuteric oxidation produces a hardening in magnetic properties at both the transition from the monophasic to the intergrown state and at the exsolution/pseudomorphic replacement boundary. In large grain single crystal $x = 0.6$ (Tucker and O'Reilly, 1980*b*), the susceptibility in the three zones is about 3, 2.5 and 0.5 respectively; the coercive force 500 A/m, 2000 A/m and 4000 A/m and M_{rs}/M_s 0.01, 0.05 and 0.5. The properties of the magnetite in the post-exsolution stage are consistent with the monodomain state. The presence of two phases in the exsolution stage results in peaks in the temperature dependence in coercive force because the higher Curie point phase is magnetically the harder.

7.3.3 *Weak field remanence*

The TRM acquired by deuterically oxidized titanomagnetite increases in intensity and stability with increasing degree of oxidation. The weak field TRMs in the two-phase region can also have the property of self-reversal (Petersen and Bleil, 1973; Tucker and O'Reilly, 1980*a*).

The specific weak field volume TRM per unit field in the monophasic region for large grain titanomagnetite is typically 0.2 (Fig. 7.17). After deuteric oxidation into the two-phase region, the specific TRM becomes about 1, and in the post-exsolution region about 5 (Tucker and O'Reilly, 1980*b*). If the latter TRM is due to a few volume percent of magnetite this indicates the magnetite is in the finely divided form ($<1\ \mu$m) (Fig. 7.16). The MDFs rise from about 10^3 A/m for the unoxidized material to about 3×10^4 A/m in the final stages.

The reversed TRM occurs in the compositional zone $z = 0.2$–0.5 (Tucker and O'Reilly, 1980*c*) and thermal demagnetization of the TRM reveals the separate reversed and normal components. A reversed moment at room temperature is not obtained for inducing fields greater than about 50 A/m. PTRMs induced in the lower Curie point phase are normal, and a PTRM induced only in the upper Curie point phase results in an enhanced reversed remanence in the assemblage at room temperature. A magnetostatic interaction between the phases is evidently the mechanism underlying the phenomenon.

Although deuteric alteration is widespread in nature, self-reversal is rarely reported, reversed remanences invariably being associated with reversals in the geomagnetic field. One, disputed, instance is the reversed NRM of volcanic rocks near Laschamp and Olby (Chaine de Puys, Auvergne, France), believed to indicate the most recent excursion ($\sim$40 000 years b.p.) of the geomagnetic field. Self-reversal due to magnetostatic interactions between titanomagnetite constituents of an intergrown oxidation product may, however, be the origin of this reversed NRM (Heller and Petersen, 1982).

7.4 Summary

The intrinsic properties of the titanomagnetites $Fe_{3-x}Ti_xO_4$ $(0 < x < 1)$ reflect the composition and the distribution of cations between the inequivalent sublattices of the spinel structure. Both saturation magnetization, M_s, and Curie temperature, T_c, fall as x increases. Observed variations in M_s for materials of the same composition may indicate that a definitive cation distribution does not exist, or that variations in high temperature non-stoichiometry influence the magnetic properties. The latter seems to be responsible for the variations in T_c in materials of nominally the same composition. The magnetocrystalline anisotropy constant K_1 and magnetostriction constants λ also depend on composition and cation distribution. At room temperature the ratio K_1/M_s increases as x (and Fe^{2+} concentration) increases, reaching about twice the magnetite $(x = 0)$ value before falling to zero at about $x = 0.8$, where T_c falls below room temperature. The ratio $\bar{\lambda}/M_s$ increases by a factor of about 20 at room temperature as x changes from 0 to about 0.6, again probably due to the increase in Fe^{2+} concentration.

The magnetization process parameters are functions of both composition and microstructure, the latter principally reflected in the grain size dependence of magnetic properties. Scatter in the experimental data for nominally similar samples indicates the difficulty in accurately specifying and quantifying microstructural parameters. At room temperature, magnetic hardness tends to increase with x and decrease with particle dimension L. The hardest materials ($x \sim 0.6$, $L < 1\ \mu m$) have $H_c > 10^5$ A/m, $M_{rs}/M_s \gtrsim 0.5$, $H_{rc}/H_c \sim 1$ and $\chi \sim 0.2$, the softest $H_c \sim 500$ A/m, $M_{rs}/M_s \sim 0.01$, $H_{rc}/H_c > 10$ and $\chi \sim 3$. The fine particle data can be modelled in terms of monodomain grains with uniaxial anisotropy, magnetostriction-based in the case of compositions with higher x. Domain walls have been observed in materials with particle dimension greater than a few μm. No definite monodomain/multidomain boundary probably exists because of microstructural variations from grain to grain but the boundary region probably lies between about 0.5 and a few μm.

The intensity and stability of TRM rise steadily as grain size falls. In fine particles ($\sim$0.1 μm) the specific volume TRM per unit field is $\sim$100, in large particles ~ 0.1, with MDFs of ~ 10^5 A/m and ~ 10^3 A/m respectively. ARM seems to exhibit an additional concentration dependence.

The saturation magnetization falls as the composition changes due to maghemitization, and the Curie point temperature rises. The variations between published contours of constant Curie temperatures in the titanomaghemite quadrilateral of the ternary diagram may be explicable in terms of experimental noise. However, the 'contour problem' may also have a systematic origin due to stoichiometry-related variations in the Curie temperatures of the titanomagnetites forming the starting point in titanomaghemite synthesis.

In the fine particle (0.02–0.05 μm) titanomaghemite in which the magnetization process has been studied, at room temperature magnetic hardness rises at first with increasing oxidation parameter z (and decreasing Fe^{2+}) before falling as z passes beyond about 0.4. The magnetization process parameters are explicable in terms of monodomain grains with uniaxial Fe^{2+}-derived anisotropy. The initial rise in hardness at room temperature occurs because Curie temperature is rising.

Metastable titanomaghemites undergo structural inversion on heating above about 350°C. This produces characteristic irreversible thermomagnetic (M_s-T) curves.

TRM can be acquired by titanomaghemites for which blocking temperatures are lower than the inversion temperature. The specific volume TRM falls approximately hyperbolically with degree of maghemitization, highly maghemitized compositions having a TRM intensity 10–20% of the equivalent unoxidized titanomagnetite, and with enhanced stability.

High temperature, deuteric, oxidation of titanomagnetite beyond the region of high temperature non-stoichiometry produces an intergrowth of phases. During the 'exsolution stage' ($0.1 < z < 0.6$ approximately) of oxidation of titanomagnetite $x = 0.6$, two titanomagnetite phases with different Curie points are present, which is some cases interact magnetostatically to produce self-reversed weak field TRM. Deuteric oxidation increases magnetic hardness especially in the post-exsolution stages when specific TRM intensity (for the intergrowth as a whole) rises to about 5 with MDF ~ 10^4 A/m. The properties of the highly oxidized intergrowth can be attributed to a dispersion of fine grain monodomain magnetite particles.

Bibliography

Akimoto, S. (1962) Magnetic properties of FeO–Fe_2O_3–TiO_2 system as a basis of rock magnetism. *J. Phys. Soc. Japan* **17**, (Suppl. B1), 706–710.

Banerjee, S.K., O'Reilly, W., Gibb, T.C. and Greenwood, N.N. (1967) Behaviour of ferrous ions in iron-titanium spinels. *J. Phys. Chem. Solids* **28**, 1328–1335.

Barton, C.E., McElhinny, M.W. and Edwards, D.J. (1980) Laboratory studies of depositional DRM. *Geophys. J.R. Astr. Soc.* **61**, 355–377.

Blasse, G. (1964) Crystal chemistry and some magnetic properties of mixed metal oxides with spinel structure. *Philips Res. Repts.* Suppl. No. 3.

Bleil, U. (1971) Cation distribution in titanomagnetites. *Z. Geophys.* **37**, 305–319.

Butler, R.F. and Banerjee, S.K. (1975) Theoretical single-domain grain size range in magnetite and titanomagnetite. *J. Geophys. Res.* **80**, 4049–4058.

Clark, D.A. and Schmidt, C.W. (1982) Theoretical analysis of thermomagnetic properties, low temperature hysteresis and domain structure of titanomagnetites. *Phys. Earth Planet. Int.* **30**, 300–316.

Clauter, D. and Schmidt, V.A. (1981) Shifts in the blocking temperature spectra for magnetite powders as a function of grain size and applied field. *Phys. Earth Planet. Int.* **26**, 81–92.

Day, R. (1977) TRM and its variation with grain size. *J. Geomag. Geoelectr.* **29**, 233–265.

Day, R., Fuller, M. and Schmidt, V.A. (1977) Hysteresis properties of titanomagnetites: grain size and compositional dependence. *Phys. Earth Planet. Int.* **13**, 260–267.

Dunlop, D.J. (1973*a*) Superparamagnetic and single domain threshold sizes in magnetite. *J. Geophys. Res.* **78**, 1780–1793.

Dunlop, D.J. (1973*b*) Thermoremanent magnetization in submicroscopic magnetite. *J. Geophys. Res.* **78**, 7602–7613.

Dunlop, D.J. (1974) Thermal enhancement of magnetic susceptibility. *J. Geophys.* **40**, 439–451.

Dunlop, D.J. (1981) The rock magnetism of fine particles. *Phys. Earth Planet. Int.* **26**, 1–26.

Dunlop, D.J. and Bina, M-M. (1977) The coercive force spectrum of magnetite at high temperatures: evidence for thermal activation below the blocking temperature. *Geophys. J.R. Astr. Soc.* **51**, 121–147.

Facey, D. (1983) A comparison between ARM and TRM in synthetic multidomain titanomagnetites. *Geophys. J.R. Astr. Soc.* **73**, 307 (Abstract).

Fletcher, E.J. and O'Reilly, W. (1974) Contribution of Fe^{2+} ions to the magnetocrystalline anisotropy constant K_1 of $Fe_{3-x}Ti_xO_4$ $(0 < x < 0.1)$. *J. Phys. C.* **7**, 171–178.

Fox, J.M.W. and Aitken, M.J. (1980) Cooling rate dependence of thermoremanent magnetization, *Nature* **283**, 462–463.

Graham, J.W. (1953) Changes in ferromagnetic minerals and their bearing on magnetic properties of rocks. *J. Geophys. Res.* **58**, 243–260.

Haggerty, S.E. (1976) 'Oxidation of opaque mineral oxides in basalts', in *Oxide Minerals* (ed. D. Rumble III), Mineralogical Society of America, Washington, D.C., Chapter 4.

Halgedahl, S. and Fuller, M. (1980) Magnetic domain observations of nucleation processes in fine particles of intermediate titanomagnetite. *Nature* **288**, 70–72.

Hauptman, Z. (1974) High temperature oxidation, range of non-stoichiometry and Curie point variation of cation deficient titanomagnetite $Fe_{2.4}Ti_{0.6}O_{4+\gamma}$. *Geophys. J.R. astr. Soc.* **38**, 29–47.

Heller, F. and Peterson, N. (1982) Self-reversal explanation for the Laschamp/Olby geomagnetic field excursion. *Phys. Earth Planet. Int.* **30**, 358–372.

Jensen, S.D. and Shive, P.N. (1973) Cation distribution in sintered titanomagnetites. *J. Geophys. Res.* **78**, 8474–8480.

Keefer, C.M. and Shive, P.N. (1981) Curie temperature and lattice constant reference contours for synthetic titanomaghemites. *J. Geophys. Res.* **86**, 987–998.

Klapel, G.D. and Shive, P.N. (1974) High temperature magnetostriction of magnetite. *J. Geophys. Res.* **79**, 2629–2633.

Kushiro, I. (1960) $\gamma \rightarrow \alpha$ transition in Fe_2O_3 with pressure. *J. Geomag. Geoelectr.* **11**, 148–151.

Larson, E., Ozima, M., Ozima, M., Nagata, T. and Strangway, D. (1969) Stability of remanent magnetization of rocks. *Geophys. J.R. astr. Soc.* **17**, 263–292.

Levi, S. (1974) Some magnetic properties of magnetite as a function of grain size and their implications for palaeomagnetism. Ph.D. thesis, University of Washington.

Levi, S. (1979) The additivity of partial thermal remanent magnetization in magnetite. *Geophys. J.R. astr. Soc.* **59**, 205–218.

Levi, S. and Merrill, R.T. (1976) A comparison of ARM and TRM in magnetite. *Earth Planet. Sci. Letts.* **32**, 171–184.

Lowrie, W. and Fuller, M. (1971) On the alternating field demagnetization characteristics of multidomain thermoremanent magnetization in magnetite. *J. Geophys. Res.* **76**, 6339–6349.

Manson, A. J., O'Donovan, J.B. and O'Reilly, W. (1979) Magnetic rotational hysteresis loss in titanomagnetites and titanomaghemites—application to non-destructive mineral identification in basalts. *J. Geophys.* **46**, 185–199.

Nagata, T., Kobayashi, K. and Fuller, M.D. (1964) Identification of magnetite and haematite in rocks by magnetic observation at low temperature. *J. Geophys. Res.* **69**, 2111–2120.

Nishitani, T. and Kono, M. (1983) Curie temperature and lattice constant of oxidized titanomagnetite. *Geophys. J.R. astr. Soc.* **74**, 585–600.
O'Donovan, J.B. and O'Reilly, W. (1980) The temperature dependent cation distribution in titanomagnetites: an experimental test. *Phys. Chem. Minerals* **5**, 235–243.
O'Donovan, J.B. and O'Reilly, W. (1982) An experimental study of TRM in Pseudo-single-domain titanomagnetite. *Geophys. J.R. Astr. Soc.* **69**, 290 (Abstract).
Ono, K., Chandler, L. and Ito, A. (1968) Mössbauer study of the ulvöspinel, Fe_2TiO_4. *J. Phys. Soc. Japan* **25**, 174–176.
O'Reilly, W. (1983) The identification of titanomaghemites: model mechanisms for the maghemitization and inversion processes and their magnetic consequences. *Phys. Earth Planet. Int.* **31**, 65–76
O'Reilly, W. and Banerjee, S.K. (1965) Cation distribution in titanomagnetites. *Phys. Letts.* **17**, 237–238.
Özdemir, Ö. and Banerjee, S.K. (1981) An experimental study of magnetic viscosity in synthetic monodomain titanomaghemites: implications for the magnetization of the ocean crust. *J. Geophys. Res.* **86**, 11, 864–68.
Özdemir, Ö. and O'Reilly, W. (1981*a*) High temperature hysteresis and other magnetic properties of synthetic monodomain titatomagnetites. *Phys. Earth Planet. Int.* **25**, 406–418.
Özdemir, Ö. and O'Reilly, W. (1981*b*) Laboratory synthesis of aluminium-substituted titanomaghemites and their characteristic properties. *J. Geophys.* **49**, 93–100.
Özdemir, Ö. and O'Reilly, W. (1982*a*) An experimental study of the intensity and stability of thermoremanent magnetization acquired by synthetic monodomain titanomagnetite substituted by aluminium. *Geophys. J.R. Astr. Soc.* **70**, 141–154.
Özdemir, Ö. and O'Reilly, W. (1982*b*) Magnetic hysteresis properties of synthetic monodomain titanomaghemites. *Earth Planet. Sci. Letts.* **57**, 437–447.
Özdemir, Ö. and O'Reilly, W. (1983) An experimental study of thermoremanent magnetization acquired by synthetic monodomain titanomaghemites. *J. Geomag. Geoelectr.* **34**, 467–478.
Parry, L.G. (1965) Magnetic properties of dispersed magnetite powders. *Philos. Mag.* **11**, 303–312.
Petersen, N. and Bleil, U. (1973) Self-reversal of remanent magnetization in synthetic titanomagnetites. *Z. Geophys.* **39**, 965–977.
Robins, B.W. (1972) Remanent magnetization in spinel iron oxides. Ph.D. Thesis, University of New South Wales.
Readman, P.W. and O'Reilly, W. (1972) Magnetic properties of oxidized (cation-deficient) titanomagnetites $(Fe, Ti, \square)_3O_4$. *J. Geomag. Geoelectr.* **24**, 69–90.
Roquet, J. (1954) Sur les remanences des oxydes de fer et leur interêt en géomagnetisme. *Ann. Geophys.* **10**, 226–247, 282–325.
Rahman, A.A., Duncan, A.D. and Parry, L.G. (1973) Magnetization of multidomain magnetite particles. *Riv. Ital. Geofis.* **22**, 259–266.
Schmidbauer, E. and Veitch, R.J. (1980) Anhysteretic remanent magnetization of small multidomain Fe_3O_4 particles dispersed in various concentrations in a non-magnetic matrix. *J. Geophys.* **48**, 148–152.
Schult, A. (1970) Effect of pressure on the Curie temperature of titanomagnetites $(1-x)Fe_3O_4xFe_2Tio_4$. *Earth Planet. Sci. Letts.* **10**, 81–86.
Soffel, H.C. and Appel, E. (1982) Domain structure of small synthetic titanomagnetite particles and experiments with IRM and TRM. *Phys. Earth Planet. Int.* **30**, 348–355.
Stephenson, A. (1969) The temperature dependent cation distribution in titanomagnetite. *Geophys. J.R. Astr. Soc.* **18**, 199–210.
Stephenson, A. (1972) Spontaneous magnetization curves and Curie points of spinels containing two types of magnetic ion. *Philos. Mag.* **25**, 1213–1232.
Sugiura, N. (1979) ARM, TRM and magnetic interactions: concentration dependence. *Earth Planet. Sci. Letts.* **42**, 451–455.
Syono, Y. (1965) Magnetocrystalline anisotropy and magnetostriction of Fe_3O_4–Fe_2TiO_4 series—with special application to rock magnetism. *Japan. J. Geophys.* **4**, 71–143.
Tucker, P. (1981) Low temperature magnetic hysteresis properties of multidomain single-crystal titanomagnetite. *Earth Planet. Sci. Letts.* **54**, 167–172.
Tucker, P. and O'Reilly, W. (1978) A magnetic study of single crystal titanomagnetite $(Fe_{2.4}Ti_{0.6}O_4)$. *Phys. Earth Planet. Int.* **16**, 183–189.

Tucker, P. and O'Reilly, W. (1980*a*) The acquisition of thermoremanent magnetization by multidomain single crystal titanomagnetite. *Geophys. J.R. Astr. Soc.* **60**, 21–36.

Tucker, P. and O'Reilly, W. (1980*b*) The laboratory simulation of deuteric oxidation of titanomagnetites: Effect on magnetic properties and stability of thermoremanence. *Phys. Earth Planet. Int.* **23**, 112–133.

Tucker, P. and O'Reilly, W. (1980*c*) Reversed thermoremanent magnetization in synthetic titanomagnetites as a consequence of high temperature oxidation. *J. Geomag. Geoelectr.* **32**, 341–355.

8 Magnetic properties of other mineral systems

8.1 Haematite

Haematite is a much studied and little understood material, and to extract what might be considered the intrinsic magnetic properties, relevant to its role as a remanence carrier, from the body of available data is problematical. It seems that microstructure and extrinsic influences affect more than just the magnetization process parameters in the case of haematite. One problem is that ideal haematite has only a weak magnetization, about 0.5 percent of that of magnetite, and the measured properties of haematite samples are correspondingly susceptible to (say) the distorting effect of undetectable quantities of magnetic contaminants. Studies on highly specified crystals have tended to concentrate on the antiferromagnetic properties, and such experiments are largely irrelevant to the fine-grain remanence-carrying haematite of rock samples. Studies of fine particle samples, prepared by crushing larger crystals, by oxidizing magnetite or by the decomposition or dehydration of a suitable iron-bearing compound, produce a range of values of saturation magnetization (after high field susceptibility has been allowed for). The magnetization also is found to fall after high temperature treatment, in air, of some of the studied material. An absolute prerequisite for an interpretable study of the intrinsic properties or magnetization process in haematite is an exhaustive characterization of the sample material. However, it may be that magnetic properties are more sensitive to (say) the presence of small quantities of maghemite than are other analytical techniques, and only indirect assessment of the material may be possible.

8.1.1 *Intrinsic properties*

At room temperature and above, the high field specific mass magnetization of haematite is expressed as $\sigma_0 + \chi H$. At room temperature σ_0, the intrinsic 'weak ferromagnetism' is $\sim 0.4\ \mathrm{Am^2/kg}$ and χ, the antiferromagnetic mass susceptibility, $25 \times 10^{-8}\ \mathrm{m^3/kg}$. Values of σ_0 between 0.2 and $0.5\ \mathrm{Am^2/kg}$ are reported, the variations ascribed to the presence of a 'defect moment'. The spontaneous magnetization falls to zero at a Curie point temperature of 675°C.

Single crystal studies show that the ferromagnetism lies in the basal plane of the corundum structure (Fig. 2.10). Neutron diffraction studies show that the spin vectors of the Fe^{3+} ions also lie in the basal plane at room temperature. This corresponds to a negative uniaxial anisotropy constant (section 3.4.4). The spin structure is antiferromagnetic, the magnetic sublattices being made up of alternating basal plane layers of Fe^{3+} ions (Fig. 2.9 and 2.10). Within each layer the Fe^{3+} spins are parallel and orientated in some direction in the basal plane. Nearest neighbour cations in adjacent layers, linked by the contracted anion triangles, are coupled antiparallel. The basal plane moments therefore alternate in direction on passing along a direction parallel to the trigonal axis.

At low temperature the orientation of the antiferromagnetic spin structure changes, the axis of the spin structure lying parallel to the trigonal axis. The uniaxial anisotropy constant is now positive. The change in sign of the anisotropy constant is interpreted in terms of competing dipolar and single ion contributions to the anisotropy (Artman and others, 1965), and the compensation point in the anisotropy, at which the spin structure changes sign, is the Morin transition (Fig. 3.11). This lies at about $-10°C$ in pure bulk haematite. Whether the Morin transition and the antiferromagnetic anisotropy should be classed as intrinsic properties of haematite is doubtful, because the temperature at which the transition takes place depends not only on composition but also on grain size. The transition temperature falls as grain size is reduced (presumably due to modification of one or both of the anisotropy contributions), although the grain-size dependence of the transition temperature differs in sample suites of differing provenance (Yamamoto, 1968). In fine enough particles ($\lesssim$ a few hundred Å) the transition is depressed below 100°K and may indeed not occur at any temperature. The introduction of other species of cations usually lowers the transition, 0.3 mole% of Ti^{4+} suppressing it completely (Besser and others, 1967).

The weak ferromagnetism associated with the high temperature (basal plane) spin structure disappears when the spin axis transforms at low temperature to the trigonal axis. Below the Morin transition, haematite is only an antiferromagnet, although not a simple one. The anisotropy of the antiferromagnetic structure can be studied by field-induced spin flopping (section 3.6.2) in which a large enough field (up to $\sim 6 \times 10^6$ A/m depending on temperature) applied parallel to the zero field spin axis, rotates the spin axis through 90°. In the case of haematite, application of a critical field normal to the spin axis produces a weak ferromagnetic state. Above the Morin transition, the spin structure can remain in the basal plane and lie normal to an applied field of any orientation, so such spin flopping experiments cannot be carried out.

The weak ferromagnetism of the basal plane spin structure arises from a non-collinear or canted spin arrangement. The spins within each cation layer are parallel to each other and to the basal plane, but the spins of adjacent planes are not exactly antiparallel. This produces a resultant moment, also in the basal plane, the magnitude of which depends on the canting angle. The size of the angle can be readily estimated from the ideal sublattice magnetizations of 175 Am^2/kg and the resultant 0.5 Am^2/kg. The canting is believed to have its origin in the combination of an antisymmetric exchange interaction and the symmetry of the corundum structure. The antisymmetric exchange interaction is represented by a vector **D** parallel to the trigonal axis (see Fig. 8.3(*a*)). The free energy may be written (Jacobs and others, 1971)

$$F = |J|\mathbf{M}_1.\mathbf{M}_2 - \mathbf{D}.\mathbf{M}_1 \times \mathbf{M}_2 + K_{u_1}(\sin^2\theta_1 + \sin^2\theta_2) + K_{u_2}(\sin^4\theta_1 + \sin^4\theta_2) - \mu_0\mathbf{H}.(\mathbf{M}_1 + \mathbf{M}_2)$$

The first term represents the (isotropic) exchange interaction. J is the exchange constant expressed in appropriate units, and $\mathbf{M}_1$ and $\mathbf{M}_2$ are the sublattice magnetizations ($|\mathbf{M}_1| = |\mathbf{M}_2| = M_0$). The second term represents the antisymmetric exchange interaction. The third and fourth terms are the crystal anisotropy energy, where θ_1 and θ_2 are the angles between the trigonal axis and the sublattice magnetizations. The final term is the magnetostatic energy in an external field. At room temperature, with the spins normal to the trigonal axis, $\theta_1 = \theta_2 = 90°$, and the anisotropy terms are constant, independent of the orientation of $\mathbf{M}_1$ and $\mathbf{M}_2$. The energy of the system is lowered by $\mathbf{M}_1$ and $\mathbf{M}_2$ lying at some other angle than 180° (Fig. 8.3(*a*)), as is required by the isotropic exchange term. It can be shown that, for $|\mathbf{J}| \gg |\mathbf{D}|$, the macroscopic magnetization (of a monodomain particle or a multidomain particle in high fields) perpendicular to the trigonal axis, for H perpendicular to this axis, is given by $DM_0/J + (\mu_0/J)H$. The spontaneous magnetization results from the balance between the antiferromagnetic and antisymmetric exchange interactions, and the susceptibility is the $\chi_\perp$ of an antiferromagnet. Thus the magnetization curve can be expressed $\sigma_H = \sigma_\perp + \chi_\perp H$. The application of a field along the trigonal axis rotates the sublattice magnetizations out of the basal plane, and the second term in the free energy is progressively reduced, as is $\sigma_\perp$. No spontaneous magnetization $\sigma_\parallel$ is possible and the magnetization curve for H parallel to the trigonal axis is $\sigma_H = \chi_\perp H$ (for exchange energy constant $\gg$ anisotropy energy constant). The measurements of the directional dependence of saturation magnetization and susceptibility of a haematite single crystal essentially support such a model (Néel and Pauthenet, 1952).

The intrinsic in-plane magnetocrystalline anisotropy constant K_3 (section 3.4.4) has been determined by Flanders and Schuele (1964) and Besser and others (1967). The small intrinsic six-fold anisotropy ($K_3 \sim 0.05$ to 20 J/m^3, several orders of magnitude less than the K_1 of Fe^{3+}- or Fe^{2+}-bearing spinel oxides) is difficult to detect in practice because of the presence of induced anisotropy. The magnetostriction coefficient at room temperature is $\sim 10^{-5}$ (Urquhart and Goldman, 1956) more or less typical of Fe^{3+}-

containing oxides. The microscopic origin of the coexistent very small (basal plane) anisotropy and unremarkable magnetostriction coefficient has not been widely discussed. Equivalent data for hexagonal oxide ferrites containing only Fe^{3+} as magnetic species and with magnetization in the basal plane are not easily found in the literature, and it is not clear whether small triaxial crystal anisotropy and larger uniaxial linear magnetostriction are inherent to the symmetry of the crystal structure or are a special property of haematite.

Above room temperature the fall in K_3 can be fitted to a power law $M_s^p(T)$ where $p = 10 \pm 1$ (Flanders and Schuele, 1964). Crystals which do not exhibit a triaxial basal plane anisotropy have the same temperature variation of spontaneous magnetization. The sublatice magnetizations (as determined from Mössbauer effect spectra) do not show this same temperature dependence and it may be inferred that the canting angle between the sublattices becomes larger as temperature rises (Flanders and Schuele, 1964).

The nature of the 'defect moment' and, indeed, a precise definition of what constitutes such a moment, have yet to be worked out. A spontaneous magnetization persisting below about $-10°C$, the Morin transition temperature in large grains, has been referred to as a defect moment. Such a moment could, of course, be intrinsic to the spin structure of material for which the Morin transition occurred at temperatures lower than $-10°C$, either due to reduced grain size or the presence of non-Fe^{3+} cations. The change in spontaneous magnetization resulting from heat treatment of fine grain synthetic αFe_2O_3 pigments is ascribed to the 'annealing out' of a defect moment (Dunlop, 1971). The presence of small quantities of a second phase, γFe_2O_3 or Fe_3O_4, which would be removed by inversion or oxidation during heating in air could account for such observations. Experiments in which the defect moment was induced rather than removed, or in which removal was accompanied by evaluation of changes in composition and microstructure by independent non-magnetic methods, would lead to a clearer understanding of the defect moment. The hypothesized mechanisms by which the defect moment may arise are discussed by Fuller (1970).

8.1.2 *The magnetization process*

At room temperature the coercive force of crystals a few mm in size is the order of 10^2–10^4 A/m (e.g. Flanders and Schuele, 1964; Eaton and Morrish, 1969), and $M_{rs}/M_s \sim 0.5$. The initial susceptibility of such soft crystals is presumably high. The high field volume susceptibility (section 8.1.1) is $\sim 1.3 \times 10^{-3}$($\sim 25 \times 10^{-8}$ m^3/kg).

The properties of materials in the grain size range 1–300 μm prepared by crushing single crystals have been determined by Chevallier and Mathieu (1943). Their figure showing the descending branches of the hysteresis loops of the crushed crystals can be found reproduced in Fuller (1970) or

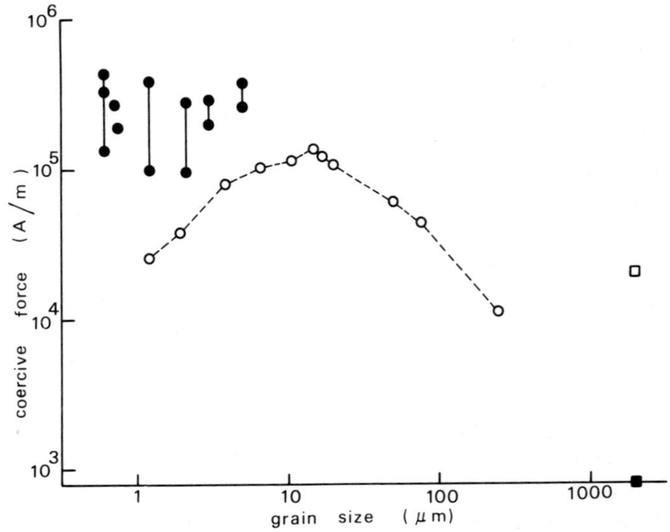

Figure 8.1 Coercive force as a function of grain size for haematite. Open circles represent data obtained by crushing natural single crystals (Chevallier and Mathieu, 1943). The finer grain material exhibits a progressively smaller ferromagnetism whereas the paramagnetic susceptibility remains the same throughout, hence the fall in coercive force as grain size falls. This characteristic is not shared by synthetic pigment material (solid circles) of similar, or smaller, size (Dunlop, 1971). The higher of the pairs of data points is the result of heating the pigment in air at 950°C. The 2 mm (nominal size) single crystal data are from Flanders and Schuele (1964) and Eaton and Morrish (1969).

Creer and others (1975), and Banerjee (1971) has plotted out the resultant grain size dependence of coercive force (Fig. 8.1). The maximum value of 1.4×10^5 A/m occurs at grain size 5–10 μm. The variation of magnetization process parameters with size, in this suite of specimens, reveals more than just the expected influence of extrinsic factors on the magnetization process. The spontaneous magnetization itself falls from about 0.4 Am^2/kg at 100 μm to a very small value at 1 μm, so that in the fine particles the induced magnetization is effectively that due to the susceptibility ($\sim 25 \times 10^{-8}$ m^3/kg) which seems characteristic of haematite. The fall in H_c cannot then be simply interpreted in terms of the magnetization process models of Chapter 4. The coercive force of remanence of sized natural haematite is variable—$(1–3) \times 10^5$ A/m for 10 μm particles—and tends to rise as grain size falls (Dankers, 1981).

The magnetization process in commercial pigment αFe_2O_3 material produced by decomposition of goethite, ferrous sulphate, oxidation of Fe_3O_4, inversion of γFe_2O_3 or 'unknown' methods, has been studied by Dunlop (1971). The grain sizes of such material lie in the range 0.1–1 μm and the presence of small quantities of magnetic contaminant is usual, and must be allowed for in inferring the properties of haematite itself. M_{rs}/M_s is about 0.5; H_c lies in the range $(1–4) \times 10^5$ A/m and H_{cr}/H_c ratios are typically

1.5–2. The high field susceptibility is the usual haematite value and the initial susceptibility ranges from about one half to three times the high field figure.

The isothermal remanence of fine particle haematite down to 0.01 μm has been determined by Hedley (1968) and Banerjee (1971), in the former case using goethite-derived material and in the latter the precipitate from an iron-containing silica gel. In both studies the remanence falls as grain size decreases although otherwise there is little quantitative agreement. Whether this fall is due only to thermal activation or arises also because of particle size dependence of 'intrinsic' properties is not clear. The presence of a high remanence ($\sim 10^{-2}$ Am^2/kg) in the finest material of Hedley's suite (~ 0.01 μm) may be an example of 'superantiferromagnetism'.

The rotational hysteresis parameters of 50 μm haematite ore grains (specularite) are $W_{rp} \sim 6$ J/kg at $H_p \sim 1.5 \times 10^5$ A/m. For synthetic pigment (~ 1 μm) the losses are 20–60 J/kg at 1.5×10^6 A/m with the probability of a higher, peak, value in higher fields. A 'non-vanishing' rotational hysteresis is believed to be intrinsic to the symmetry of the anisotropy expected in αFe_2O_3 (Owens, 1982).

As discussed in section 7.1.2 the difficulty in determining or specifying the microstructural or other extrinsic factors, which combine with the intrinsic properties in controlling the magnetization process, means that no definitive values for magnetization process parameters are to be had. In the case of haematite, an added complication arises because of the uncertainty in how truly intrinsic the intrinsic properties of spontaneous magnetization, anisotropy and magnetostriction are. Although haematite has low anisotropy and a similar magnetostriction to (say) magnetite, because of its weak magnetization, it is only weakly coupled to an external field. The coercive force can then be high given the presence of a suitable internal stress, and the values in fine grain, presumably monodomain, material can readily be accounted for by model mechanisms. Superparamagnetic/stable monodomain transition size model values may be derived using anisotropy energy barriers based on the room temperature coercive force of fine grain material. Values of several hundred Å are obtained. The modelling of the monodomain/multidomain transition has not been carried out.

The temperature dependence of the coercive force has been measured by Flanders and Schuele (1964) and Dunlop (1971). From 20°C to 500°C the coercive force of the natural single crystal studies by Flanders and Scheule was found to follow a power law $H_c \propto M_s^p$ where $p \sim 3$. The decay of coercive force with descending temperature, on passing through the Morin transition, was also reported for the same crystal. In some fine particle pigment the power indices between about 400 and 550°C and above 550°C were 8 and 2 respectively. The effect of cooling remanence-carrying haematite samples has been studied by a number of workers. When a material undergoes the Morin transition and the spontaneous magnetization vanishes, so should any associated remanence. In a typical experiment, such as those described by Smith and Fuller (1967), a sample is given an IRM at room temperature and the remanence monitored on cooling down to liquid nitrogen temperature. A large fraction of the remanence may be lost, that which remains either being due to part of the sample not undergoing the transition or due to the presence of a 'defect moment' not linked to the room temperature spin canted structure. On warming back to room temperature in zero field, a substantial recovery of remanence may take place—the so-called 'memory effect'.

The domain structure in synthetic haematite single crystals at room temperature consists of slab or cylindrical domains parallel to the basal plane separated by 180° walls (Eaton and Morrish, 1969). In thin sections of crystal (25–200 μm) the domain structure may be ill-defined. Barkhausen jumps of up to 100 μm occur in these soft crystals which saturate in a field of 2000 A/m. The walls are separated by 50–150 μm. The calculated wall energy is then of the order of 10 $\mu J/m^2$. On cooling through the Morin transition, regions without domain structure nucleate and grow until no domains remain. The converse is observed on a warming cycle.

The magnetization process parameters of fine grain haematite are variable and depend on the provenance of the material. Whether any of the data obtained, on crushed ores or fine grain synthetic material, does serve as a guide to the nature of the haematite of rocks has not been established.

8.1.3 *Weak field remanence*

The specific weak-field volume TRM intensity for fine grain material is ~0.4 (Roquet, 1954; Dunlop, 1971). A single crystal a few mm in size readily gains an intense TRM, with a specific weak field (basal plane) intensity of about 30 and a saturated value (~0.8 times the saturation moment) in a few hundred A/m (Syono and others, 1962). The stabilities of the TRMs reflect the coercive forces of the materials. Thus the MDF of fine grain material exceeds the normally available laboratory AF ($\sim 2 \times 10^5$ A/m) whereas the single crystal 800 A/m TRM had an MDF of about 10^4 A/m.

The acquisition of CRM by haematite during its formation by dehydration of the four isomers of ferric oxyhydroxide has been studied by Hedley (1968). The specific weak field volume CRMs were typically 10^{-3} in the case of the dehydration of goethite (α–FeOOH) and akaganeite (β–FeOOH). The production of αFe_2O_3 from lepidocrocite (γ–FeOOH) involves γFe_2O_3 as an intermediate phase, apparently near the superparamagnetic limit at room temperature in Hedley's experiments, in which the ultimate αFe_2O_3 also showed viscous behaviour. In the case of acicular γFeOOH, the intermediate γFe_2O_3 remanence was reversed. The dehydration of (strongly magnetic) δ–FeOOH to αFe_2O_3 has α–FeOOH as an intermediate phase. The interpretation of the CRM in terms of the mechanism of section 6.1.1

is complicated when one magnetic phase transforms to another.

The DRM process operates efficiently with coarse grain haematite (specularite), the DRM acquisition curve being linear with field (and with slope ~ 2) up to a few hundred A/m (Collinson, 1974).

An acquisition coefficient, S, of 0.13 A/m has been found for the VRM acquired by 0.07–0.15 μm haematite in a field of 400 A/m (Dunlop and Stirling, 1977).

8.2 The haematite-ilmenite solid solution series

Like haematite, the haematite-ilmenite solid solution, $Fe_{2-y}Ti_yO_3$ $(0 < y < 1)$, has variable fundamental properties. A defect moment is not the origin of the variability in this quite strongly magnetic system. The equivalence of cation layers in the corundum/ilmenite structure and the possibility of a temperature dependent cation distribution with, at either extreme, an ordered or disordered $Fe^{2+} - Ti^{4+}$ distribution between the layers, has been mentioned in section 2.1.3. An ordered cation arrangement would have a net magnetization when the spin order is ferromagnetic within each layer, and alternate layers are oppositely magnetized as in haematite. The low temperature spontaneous moment would be $4y\beta$ per formula unit for a spin-only collinear system. A model moment of this intensity would not be expected to persist up to $y = 1$, however, as alternate Ti-rich layers have low or zero concentration of magnetic species with which the moments of the Fe^{2+}-rich layers can couple antiparallel. Ilmenite itself is antiferromagnetic ($T_N \sim 55°$K) with spin vectors along the c axis. A model moment for the ordered cation distribution would fall from the linear $n_\beta = 4y$ relationship as y approaches unity, although a quantitative model has not been worked out. The moment for the disordered cation distribution would be zero throughout the series for a collinear spin arrangement.

8.2.1 *Intrinsic properties*

The saturation magnetization data of Bozorth and others (1957) and Ishikawa and Akimoto (1958) are reproduced in Fig. 8.2. Some of the variation between different investigations may be due to differences in the experimental conditions of applied field and temperature, but variations no doubt arise from differences in thermal history, as the comparison between the quenched and slowly cooled samples of the same composition suggests. The values of n_β (β pfu) of Fig. 8.2 can be converted to the macroscopic units Am^2/kg by multiplying by a factor $N_0\beta/M$, where N_0 is Avogadro's number ($N_0\beta = 5585$) and M the molecular weight $(160 - 8y)$. The measured moments of slowly cooled samples approach the model ordered moment in the range $0.5 \lesssim y \lesssim 0.8$. The low values of n_β at $y < 0.4$ may correspond to a disordered cation distribution; the low values as y approaches one are

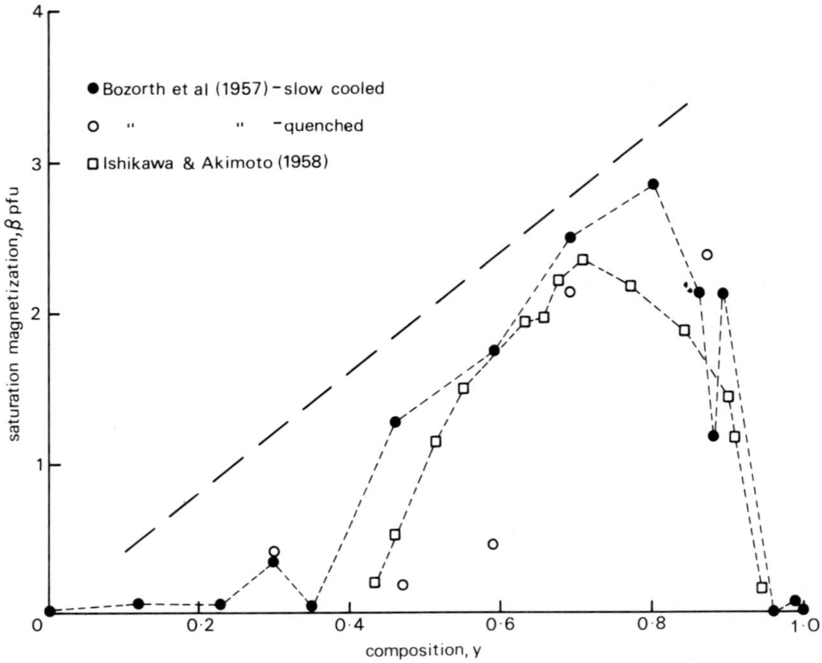

Figure 8.2 Measured low temperature saturation magnetization of the haematite-ilmenite series. The dashed line $n_\beta = 4y$ corresponds to an ordered cation arrangement and collinear spin structure. $n_\beta = 0$ would apply for a disordered cation arrangement and also for the ordered antiferromagnetic structure of ilmenite. It may be that the Fe^{3+} present in the region near $y = 1$ and the Fe^{2+} near $y = 0$ form a third magnetic sublattice. At room temperature, non-collinear canted spin structures apply near $y = 0$.

appropriate to either order or disorder. The ordered cation structure, and a simple antiferromagnetic spin structure, are inferred from neutron diffraction studies of $FeTiO_3$. The spin structures near either end of the series may be complicated. Near $y = 1$, with an ordered cation arrangement, small quantities of Fe^{3+} occur in both the magnetic Fe^{2+} layers and non-magnetic Ti^{4+} layers. If ferromagnetic order occurs in the Fe^{2+} layers and the moments of the layers are alternately aligned along the c axis, what is the orientation of the Fe^{3+}? A third (or fourth, i.e. antiferromagnetic order in Fe^{3+}, Ti^{4+} planes) sublattice may be formed or, if the Fe^{3+} is not uniformly dispersed throughout the crystal but is locally concentrated, regions with a different spin structure may occur (Ishikawa, 1962). Near the $y = 0$ end of the series the canted spin structure of haematite may exist at all temperatures (i.e. the Morin transition is suppressed) and a resultant magnetization will occur even in the disordered state. A further possibility is that the Fe^{2+} will form a third sublattice (Searle and Morrish, 1966) with magnetization opposed to the resultant of the canted Fe^{3+} moments (Fig. 8.3). The spontaneous magnetization would then be $DM_0/J + M_3(J_3/J - 1)$

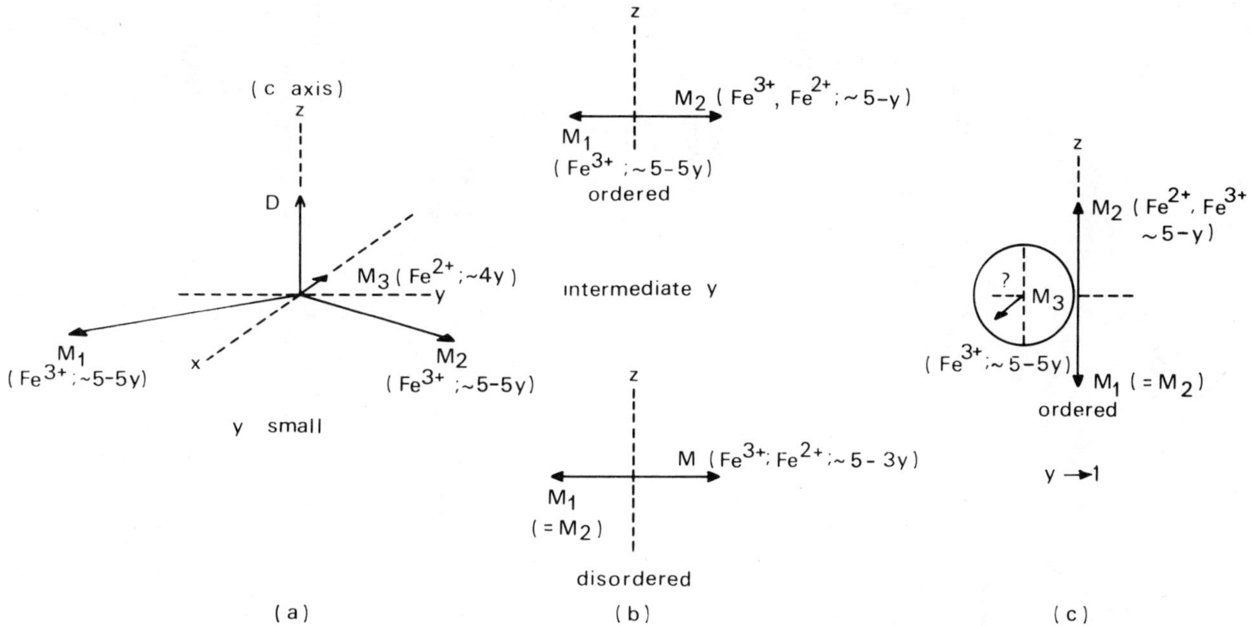

Figure 8.3 Spin structures of the haematite-ilmenite solid solution $Fe_{2-2y}Ti_yO_3$. (*a*) Near the haematite end of the series and at a temperature above the Morin transition, M_1 and M_2 are the canted Fe^{3+} moments in the basal plane. M_3 is the opposed moment of Fe^{2+} ions forming a third magnetic sublattice. The sublattice magnetizations in β pfu are shown. (*b*) Intermediate *y*, for which the sublattice magnetizations are taken to lie in the basal plane. The sublattice population and magnetization are shown for the ordered and disordered cation distributions. (*c*) Near the ilmenite end (and below room temperature, $T_N \sim 55°K$ for $y = 1$). In the ordered arrangement, alternate layers are occupied by small concentrations of Fe^{3+}. That the ions in these layers form a third sublattice is speculative.

where the notation is the same as in section 8.1.1, M_3 is the moment of the third sublattice and J_3 is proportional to the exchange energy constant for the $Fe^{2+} - Fe^{3+}$ interaction. The possible or actual spin structures in the solid solution series are indicated in Fig. 8.3. It seems that the preferred axis of the spin structure changes from basal plane (for T/T_c above the Morin transition) at $y = 0$ to c axis at $y = 1$. At some point in the series, and for constant T/T_c, the spin axis must change from plane to axis. Where this occurs may depend on temperature. In Fig. 8.3 the change in spin axis is presumed to take place near the $y = 1$ end of the series.

Quenching and annealing experiments led Ishikawa (1962) to propose that the cation distribution is not a continuous function of temperature, in which progressive disorder correlates with progressively higher temperature, but that a critical point temperature between the ordered and disordered states occurs at 800–1100° K depending on y. The quenching/annealing experiments of Hoffman (1975) do not support a simple order/disorder model.

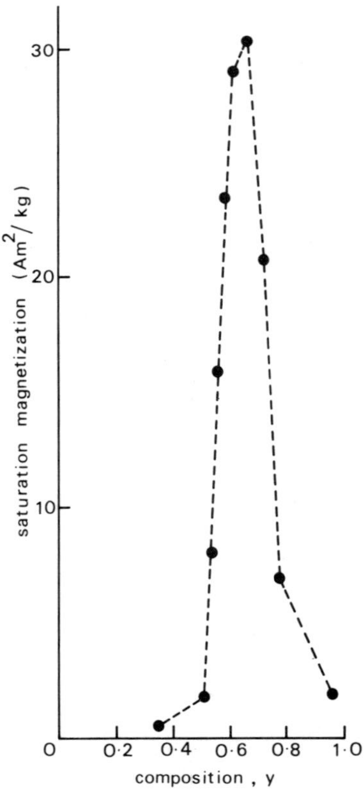

Figure 8.4 The variation of the saturation magnetization at room temperature with composition for haematite-ilmenite samples of grain size 9.8 μm.

The Curie point temperature falls steadily as y increases. The variation has been reported as $T_c(y) = 675–885y$ (°C) (Nagata and Akimoto, 1956) and $T_c(y) = 622 - 655y$ $(0.3 < y < 1)$ (Westcott-Lewis and Parry, 1971*a*). The Curie points of quenched samples are typically 50°C lower than annealed material in the range $0.45 < y < 0.8$ (Ishikawa, 1962). Whether variations in stoichiometry can exert a strong influence on the Curie point, as is the case with titanomagnetites (section 7.1.1.2), is not known.

The room temperature saturation magnetization, which may be considerably lower than (say) the liquid nitrogen temperature value, is presented by Westcott-Lewis and Parry (1971*a*) whose data for 9.8 μm diameter grains is reproduced in Fig. 8.4. Westcott-Lewis and Parry found the saturation moment (or at least the induced moment in 8×10^5 A/m) to depend on grain size, the smaller grains having smaller moment and the effect being greatest in the $y \sim 0.5–0.6$ region.

8.2.2 *The magnetization process*

The magnetization process has been studied for the composition range $0.51 \leqslant y \leqslant 0.78$ and the particle size range $1.5\ \mu m \leqslant L \leqslant 26\ \mu m$ by Westcott-Lewis and Parry (1971*a*), principally at room temperature. The room temperature parameters H_c, H_{cr}/H_c, M_{rs}/M_s and (mass) χ are plotted in Fig. 8.5 for grain size 9.8 μm. The mass susceptibility is converted to volume susceptibility by multiplying by the density (X-ray density $\sim (5.3 - 0.5y) \times 10^3$ kg/m^3). All measured properties for this grain size range show a maximum hardness in the region of $y = 0.5 - 0.7$. The change in composition takes the spontaneous magnetization through a peak at room temperature (and probably all temperatures) and reduces the Curie point temperature. How the magnetostriction and crystal anisotropy constants change with composition is not known, although we might suppose that these properties are progressively dominated by the Fe^{2+} present in $Fe^{3+}_{2-2y}Fe^{2+}_{y}Ti^{4+}_{y}0_3$. The interplay of the composition dependences of the intrinsic parameters can then account qualitatively for the composition dependence of the magnetization process parameters, microstructural or other extrinsic influences remaining constant. Thus as y increases, susceptibility peaks after saturation magnetization has fallen from its peak value, presumably because the proximity of the falling Curie point has reduced anisotropy. The coercive force peaks before the saturation magnetization does because the hardness parameter (anisotropy/M_s) starts to fall both directly due to composition change and indirectly, via the rising magnetic isotherm. Data for material of the same provenance (and similar microstructural character) for the $0 < y < 0.5$ range is not available but it might be expected that hardness increases again, after passing through a minimum, as $y = 0$ is approached. Westcott-Lewis and Parry (1971*a*) quote the haematite-like values of $M_{rs}/M_s = 0.54$ and $H_c = 9 \times 10^4$ for undispersed grains of $y = 0.35$ ($\sigma_s = 0.35$ Am^2/kg) at room temperature.

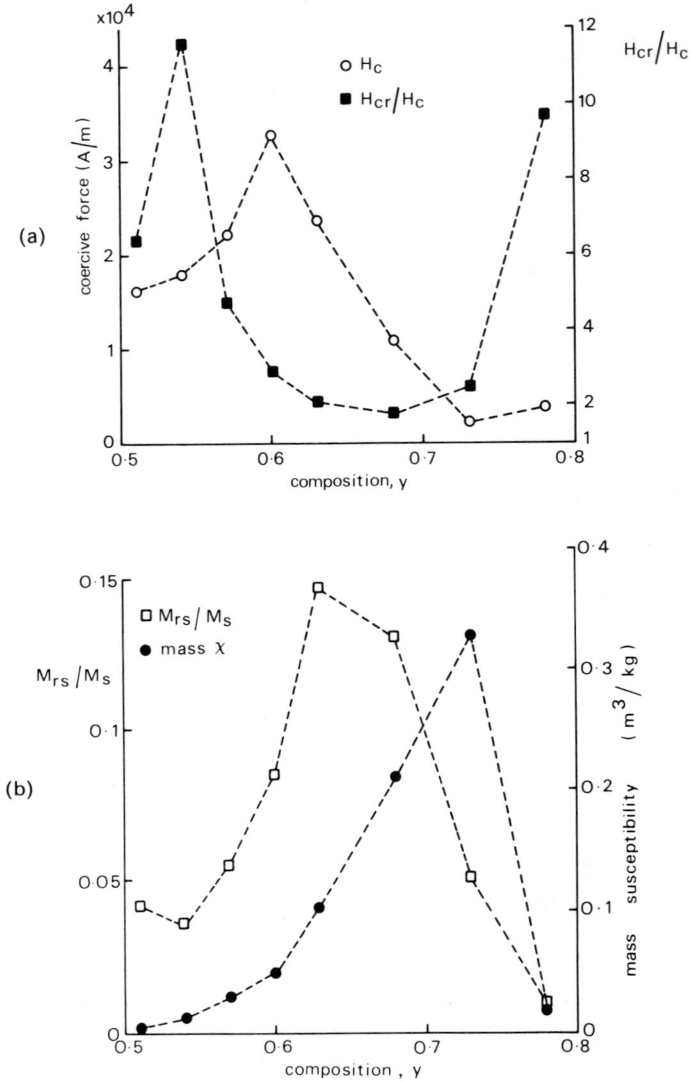

Figure 8.5 Magnetization process parameters at room temperature for 9.8 μm grains of haematite-ilmenites ($0.5 < y < 0.8$). (*a*) H_c and H_{cr}/H_c; (*b*) M_{rs}/M_s and (mass) susceptibility. (Data from Westcott-Lewis and Parry, 1971*a*).

The grain size dependence of the magnetization process parameters for $y = 0.63$ are plotted in figure 8.6. When the power laws M_{rs}, $H_c \propto L^{-n}$ are fitted to the data for this and other compositions, values of n between 0.2 and 0.5 are obtained (Westcott-Lewis and Parry, 1971*a*). At the smallest grain size studied, $M_{rs}/M_s (= 0.26)$ and $H_{cr}/H_c (= 1.8)$ have yet to reach model monodomain values.

(a)

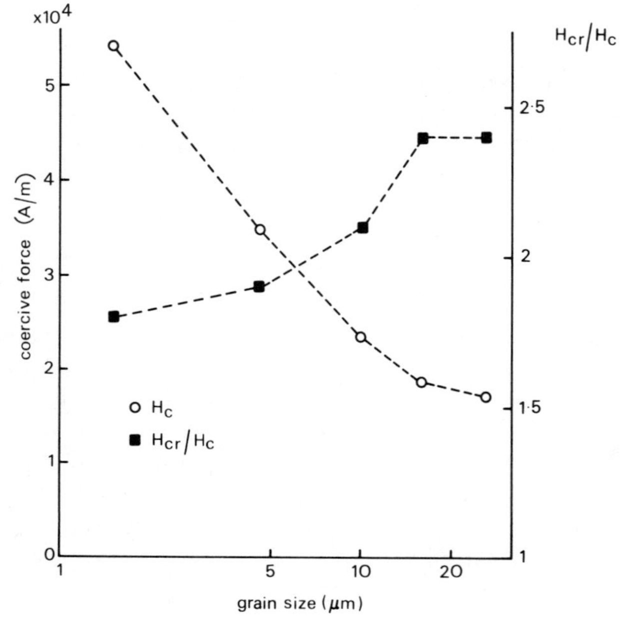

(b)

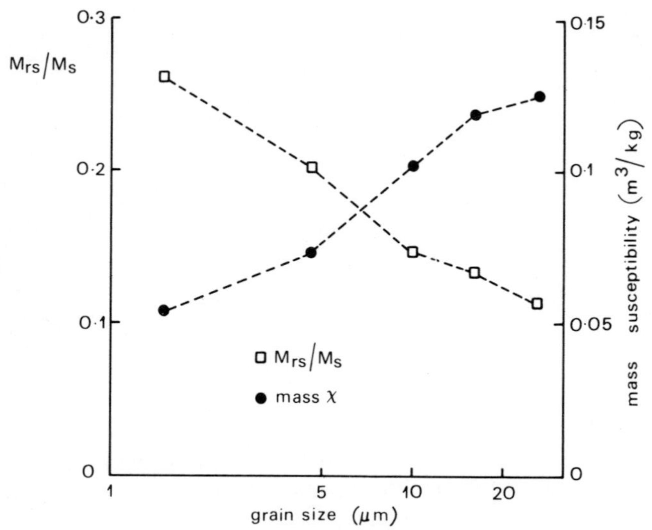

Figure 8.6 The room temperature magnetization process parameters for haematite-ilmenite of composition $y = 0.63$ in the grain size range 1–30 μm. (*a*) H_c and H_{cr}/H_c; (*b*) M_{rs}/M_s and mass susceptibility. (Data from Westcott-Lewis and Parry, 1971*a*.)

Slowly cooled natural haematite-ilmenite may undergo subsolvus exsolution. The miscibility gap ($0.25 \lesssim y \lesssim 0.75$) may be rather narrower than that of the titanomagnetites (Fig. 2.11) and the consolute temperature rather higher ($\sim 800°C$) (Lindsley, 1976). The presence of subsolvus exsolution will clearly influence the magnetization process (Merrill, 1968) and if undetected in synthetic material lead to misinterpretation of intrinsic properties.

8.2.3 *Weak field remanence*

The haematite-ilmenites are remarkable in showing reversed thermoremanence in the compositional range $0.5 < y < 0.7$. For the details of some of the experiments aimed at elucidating this phenomenon the reader is referred to the papers of Ishikawa and Syono (1963), Westcott-Lewis and Parry (1971*b*) and Hoffman (1975). A two-phase mechanism (section 5.4) seems to operate in the system, although the identity of the phases is not known. It may be that regions of order and disorder coexist with Fe-rich regions developed during the ordering process or by subsolvus exsolution. Some of the experimental facts are that: shifted hysteresis loops are observed in specimens cooled in the presence of a field; the self-reversal behaviour is affected by quenching/annealing operations; the self-reversal is limited to grains above a certain size; TRM is normal in large enough applied field.

The normal specific volume TRM of 1.5 μm grains of $y = 0.68$ is about 4 (Westcott-Lewis and Parry, 1971*b*).

8.3 The pyrrhotites

The complexity of the pyrrhotite system is indicated by the phase diagram of Fig. 8.7, a simplified picture based on the work of Schwarz and Vaughan (1972) and earlier workers. The crystal structure was discussed in section 2.1.4. At temperatures below about 310°C, zones of solid solution may be restricted to the regions near the indicated compositions. If so, intermediate compositions would consist of mixtures of these compounds. The nonstoichiometric pyrrhotites ($Fe_{1-x}S$, $x > 0$) undergo the 'β-transition' at 310–320°C, above which the vacancies are randomly distributed among the basal plane layers, and below which they form an ordered superstructure. A further structural transformation takes place at the 'γ-transition' for compounds in the range $0.08 < x < 0.11$.

8.3.1 *Intrinsic properties*

Monoclinic pyrrhotite (Fe_7S_8) is believed to have the cation vacancies located in alternate basal plane layers below the β-transition. At, or possibly slightly below the transition, magnetic ordering takes place at the Curie

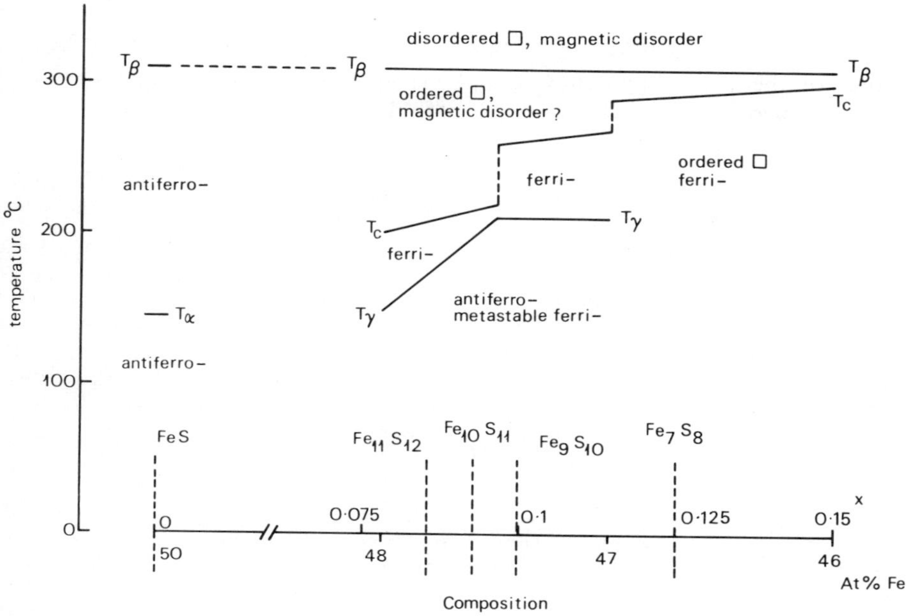

Figure 8.7 Simplified and schematic phase diagram for the troilite/pyrrhotites. The extent of solid solution in the system below the β-transition may be limited to regions near the indicated compositions.

point ($\sim 300°C$). As in haematite, the spin structure consists of ferromagnetic order within basal planes, the moments of adjacent planes being coupled antiparallel. Unlike haematite, the moments of the planes differ in magnitude and a strong ferrimagnetism results. The valency of the iron in the pyrrhotites is not definitely established, but the weight of evidence suggests that all the iron is Fe^{2+}, and that the average valency of the sulphur is less than 2 (Ward, 1970). The structural formula unit could then be written $|Fe_4^{2+}||Fe_3^{2+}\square|$ $(S_8^{2-}h_2^{\ +})$ in which the two magnetic sublattices are shown within parentheses and h^+ represents a positive hole (i.e. the absence of an electron) associated with the sulphur ions. This suggests that a fraction of the sulphur ions do not have the Ar inert gas electron structure (Table 3.1) and will carry a magnetic moment. Assuming that any moments of the sulphur layers cancel, the resultant moment due to the imbalance in cation populations is 4β pfu (32 Am^2/kg or about 1.4×10^5 A/m). Experimental values of the low temperature saturation moment of pyrrhotite crystals are lower than this. At room temperature the saturation magnetization is about 20 Am^2/kg.

At room temperature the spontaneous magnetization is coupled to the basal plane by a strong crystal anisotropy ($\gtrsim 10^5$ J/m^3). The anisotropy within the basal plane has a pronounced triaxial component. From the absence of stress patterns around scratches, noted during domain studies,

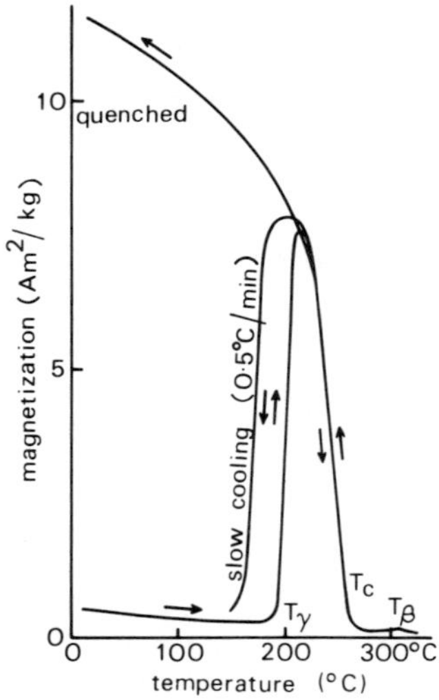

Figure 8.8 The peak type thermomagnetic curve of synthetic low temperature hexagonal pyrrhotite Fe_9S_{10}. The ferrimagnetic regime, stable between the γ-transition and the Curie point, may be preserved below the γ-transition by quenching. (Data of Schwarz, 1975.)

Soffel (1977) estimated 7×10^{-6} as an upper limit to the magnetostriction constant.

The low temperature hexagonal pyrrhotites, compositions at or regions near Fe_9S_{10}, $Fe_{10}S_{11}$ and possible $Fe_{11}S_{12}$, also exhibit ferrimagnetism between the γ-transition (at about 220°C, 210°C and 180°C respectively for the three compositions, see Fig. 8.7) and the Curie points (270°C, 230°C and 210°C respectively). Below the γ-transition, e.g. at room temperature, the stable state is antiferromagnetic. The hexagonal pyrrhotites therefore show 'peak-type' thermomagnetic curves, in which the magnetization rises rapidly to a peak value before falling to the Curie point (or at least a transition from the ferrimagnetic state). The peak-type behaviour is shown schematically in Fig. 8.8. On cooling slowly (at, say, 0.5°/min) the material reverts to its weakly or non-magnetic low temperature state, thermal hysteresis being observed in the transition. If, however, the sample is quenched from the ferrimagnetic zone, the ferrimagnetism may be preserved at room temperature. This can be removed by annealing just below the γ-transition temperature. Various hypotheses have been put forward to explain the antiferromagnetic/ferrimagnetic γ-transition (Ward, 1970). One is that the transition represents an upper limit to the region in which (say) Fe_9S_{10} can exist as a single phase and, above this limit, samples contain Fe_7S_8. Another is that transformations occur between one ordered vacancy superstructure and another, one having a balanced, and the other an imbalanced, spin structure. The moment of the quenched-in ferrimagnetism of Fe_9S_{10} is about 11 Am^2/kg in a field of a few ($\times 10^5$) A/m.

Troilite, FeS, is antiferromagnetic with susceptibility 20×10^{-8} m^3/kg at room temperature.

8.3.2 *The magnetization process*

Because of the strong anisotropy, the magnetization curve of Fe_7S_8 along the hard [0001] axis is approximately linear with field, like a paramagnet. Schwarz and Vaughan (1972) estimated the effective high field susceptibility along the hard axis to be 70×10^{-8} m^3/kg, from the room temperature magnetization curve of a polycrystalline assemblage. The weak field susceptibility in the basal plane is of the same order as the susceptibility of magnetite (Schwarz, 1975). Polycrystalline Fe_7S_8 acquires a moment of 15 Am^2/kg in a field of 4.3×10^5 A/m.

The hysteresis loop of a Devonian diabase containing predominantly pyrrhotite with a wide range of grain size, from < 1 μm to > 100 μm, is presented by Soffel (1981). The coercive force is 1.4×10^4 A/m and M_{rs}/M_s about 0.5. An M_{rs}/M_s ratio of about 0.4 has been obtained for a pyrrhotite crystal a few mm in size (Syono and others, 1962). For a sample of ore, Halgedahl and Fuller (1981) obtained $H_c = 9 \times 10^3$ A/m and $M_{rs}/M_s = 0.14$. It might be noted at this point that the model monodomain M_{rs}/M_s for the symmetry of pyrrhotite—a unique hard axis and a triaxial anisotropy in the unique plane—has been calculated as 0.750 (Dunlop, 1971). A higher value than this would be inferred due to an underestimated measured M_s, if the applied field was not great enough to saturate a random assemblage of particles.

The domain observations of Soffel (1971) place the monodomain/multidomain transitional zone at about 2 μm.

8.3.3 *Weak field remanence*

A specific weak-field basal plane TRM of 20 has been found for a pyrrhotite single crystal of dimensions a few mm (Syono and others, 1962), an 800 A/m TRM having an MDF of 5×10^3 A/m.

The non-uniqueness of domain structures is emphasized by the observations of Halgedahl and Fuller (1981) on TRM-carrying, ARM-carrying, thermally and AF demagnetized polycrystalline ore grains. In the TRM-carrying or thermally demagnetized state, sections of the undulating domain walls were easily moved in weak fields. The ARM-carrying or AF demagnetized state was characterized by a relatively simple pattern of parallel walls. The total wall length and initial susceptibility were greater in the former case.

Pyrrhotite ores are often mixtures of monoclinic and low temperature hexagonal pyrrhotite (Fe_9S_{10}) and may include some magnetite. The thermomagnetic properties of the ores are correspondingly complicated (Schwarz, 1975) not least in their remanence properties, where interactions between constituent phases can produce reversed PTRM (Bhimasankaram and Lewis, 1966).

8.4 Goethite

The spin ordering in goethite, γ-FeOOH, is antiferromagnetic with the spin axes parallel to one edge of the orthorhombic unit cell. Samples of goethite also exhibit a weak ferromagnetism (or ferrimagnetism) which, single crystal studies suggest, lies parallel to the antiferromagnetic spin axis (Hedley, 1971). The disappearance of the weak ferromagnetism coincides with the Néel point of the antiferromagnetism. Observed saturation magnetizations lie in the range 10^{-3} to 1 Am^2/kg and the Néel point temperature ranges from 60° to 170°C.

The origin of the weak ferromagnetism is unknown. Hedley (1971) suggests that an imbalance arises between the antiferromagnetic sublattices due to the preferential location of Al^{3+} on one sublattice.

Natural samples of goethite probably carry a CRM. The natural remanence is about two orders of magnitude smaller than the saturation moment. Strangway and others (1968) have observed weak and variable, but stable, TRMs in samples cooled through the Néel point.

8.5 Iron

Iron is ferromagnetic with a Curie temperature of 770°C. The saturation magnetization at room temperature is 1.7×10^6 A/m (220 Am^2/kg). The magnetocrystalline anisotropy constants are $K_1 = 4.7 \times 10^4\ J/m^3$ and $K_2 = 1.5 - 3.0 \times 10^4\ J/m^3$ at room temperature, and the magnetostriction coefficients $\lambda_{100} = 25 \times 10^{-6}$, $\lambda_{111} = -19 \times 10^{-6}$.

Bulk iron is very soft, the coercive force of purified sheet iron being a few A/m. Permanent magnets, made from iron powder, have coercive force $\sim 6 \times 10^4$ A/m and $M_{rs}/M_s \sim 0.4$. The particle size dependence of the hysteresis parameters of fine particle iron are reproduced by Dunlop (1981). M_{rs}/M_s is equal to about 0.5 in the range $0.01 < L < 0.05$ μm (at 207°K). The coercive force peaks at $H_c \sim 8 \times 10^4$ A/m, $L \sim 0.03$ μm. Shape anisotropy no doubt dominates the magnetization process in this, the most strongly magnetic substance. The fall in H_c with decreasing particle size can be fitted to the thermal fluctuation model (section 4.4.4 and Fig. 6.1) (Dunlop, 1981).

The specific volume TRM of ex-carbonyl iron is about 0.01 and the specific ARM about 0.008 (Stephenson and Collinson, 1974). The MDF of an 80 A/m TRM is about 2.5×10^3 A/m.

8.6 Alteration products of non-magnetic minerals

Non-magnetic iron-bearing phases may alter by oxidation or decomposition to produce simple oxides, and the properties of altered phases are often interpretable in terms of magnetite or haematite in a given concentration and with a given microstructure.

The low temperature oxidation of olivines to produce magnetite below the Curie point, and the acquisition of CRM, ARM and TRM by the magnetite has been described by Hoye and Evans (1975). The remanences acquired by altered material in an 800 A/m field are of the order of $10^{-2} - 10^{-3}$ Am^2 per kg of olivine with MDFs in the range $(1.6 - 5) \times 10^4$ A/m.

A TRM of 1.4×10^{-3} Am^2/kg, acquired by altered magnetite-bearing pyroxene (Evans and others, 1968), could account for a large fraction of the remanence of the gabbro from which it was extracted.

A recent example of the study of the 'magnetic enhancement' of clay minerals by heat treatment is the work of Moskowitz and Hargraves (1982) in which nontronite heated at 700 or 900°C reddens in colour and acquires a saturation magnetization of up to 4.4 Am^2/kg. Maghemite (γFe_2O_3) and haematite (αFe_2O_3) may be the magnetic alteration products. Other minerals, e.g. chamosite, can be magnetically enhanced at lower temperatures (~ 400°C).

8.7 Summary

The magnetic minerals which are not spinel oxides are relatively complicated and ill-behaved: haematite because of its weak and variable spontaneous magnetization and the effect of grain size on its 'intrinsic' anisotropy; the haematite-ilmenite solid solution because of the equivalence of all cation sites in the crystal structure, and the (inferred) coexistence of magnetically different phases, depending on thermal history; the pyrrhotites because of their complicated phase diagram and the effect of thermal history on vacancy distribution among equivalent lattice sites.

Bibliography

Artman, J.O., Murphy, J.C. and Foner, S. (1965) Magnetic anisotropy in antiferromagnetic corundum-type sesquioxides. *Phys. Rev.* **138**, A912–A917.

Banerjee, S.K. (1971) New grain size limits for the palaeomagnetic stability in haematite. *Nature Phys. Sci.* **232**, 15–16.

Besser, P.J., Morrish, A.H. and Searle, C.W. (1967) Magnetocrystalline anisotropy of pure and doped hematite. *Phys. Rev.* **153**, 632–640.

Bhimasankaram, V.L.S. and Lewis, M. (1966) Magnetic reversal phenomena in pyrrhotite. *Geophys. J.R. Astr. Soc.* **11**, 485–497.

Bozorth, R.M., Walsh, D.E. and Williams, A.J. (1957) Magnetization of haematite-ilmenite system at low temperatures. *Phys. Rev.* **108**, 157–158.

Chevallier, R. and Mathieu, S. (1943) Propriétés magnetiques des poudres d'hématites; influence des dimensions des grains. *Ann. Phys.* **18**, 258–288.

Collinson, D.W. (1974) The role of pigment and specularite in the remanent magnetization of red sandstones. *Geophys. J.R. Astr. Soc.* **38**, 253–264.

Creer, K.M., Hedley, I.G. and O'Reilly, W. (1975) 'Magnetic oxides in geomagnetism', in *Magnetic Oxides*, (ed. D.J. Craik), John Wiley & Sons, London, New York, Sydney, Toronto, Chapter 11, pp. 649–688.

Dankers, P. (1981) Relationship between median destructive field and remanent coercive forces for dispersed natural magnetite, titanomagnetite and hematite. *Geophys. J.R. Astr. Soc.* **64**, 447–461.

Dunlop, D.J. (1971) Magnetic properties of fine-particle hematite. *Ann. Géophys.* **27**, 269–293.

Dunlop, D.J. (1981) The rock magnetism of fine particles. *Phys. Earth Planet. Int.* **26**, 1–26.

Dunlop, D.J. and Stirling, J.M. (1977) "Hard" viscous remanent magnetization (VRM) in fine-grained hematite. *Geophys. Res. Letts.* **4**, 163–166.

Eaton, J.A. and Morrish, A.H. (1969) Magnetic domains in hematite at and above the Morin transition. *J. Appl. Phys.* **40**, 3180–3185.

Evans, M.E., McElhinny, M.W. and Gifford, A.C. (1968) Single domain magnetite and high coercivities in a gabbroic intrusion. *Earth Planet. Sci. Letts.* **4**, 142–146.

Flanders, P.J. and Schuele, W. (1964) 'Temperature dependent magnetic properties of hematite single crystals, in *Proceedings of the International Conference of Magnetism, Nottingham, September 1964*, Institute of Physics and the Physical Society, London, pp, 594–596.

Fuller, M.D. (1970) Geophysical aspects of paleomagnetism. *Crit. Rev. Solid State Sci.* 137–219.

Halgedahl, S.L. and Fuller, M. (1981) The dependence of magnetic domain structure upon magnetization state in polycrystalline pyrrhotite. *Phys. Earth Planet. Int.* **26**, 93–97.

Hedley, I.G. (1968) Chemical remanent magnetization in the FeOOH, Fe_2O_3 system. *Phys. Earth Planet. Int.* **1**, 103–121.

Hedley, I.G. (1971) The weak ferromagnetism of goethite (α-FeOOH). *Z. Geophys.* **37**, 409–420.

Hoffman, K.A. (1975) Cation diffusion process and self-reversal of thermoremanent magnetization in the ilmenite-haematite solid solution series. *Geophys. J.R. Astr. Soc.* **41**, 65–80.

Hoye, G.S. and Evans, M.E. (1975) Remanent magnetizations in oxidized olivine. *Geophys. J.R. Astr. Soc.* **41**, 139–151.

Ishikawa, Y. (1962) Magnetic properties of ilmenite-hematite system at low temperature. *J. Phys. Soc. Japan* **17**, 1835–1844.

Ishikawa, Y. and Akimoto, S. (1958) Magnetic property and crystal chemistry of ilmenite ($MeTiO_3$) and hematite (αFe_2O_3) system, 2: Magnetic property. *J. Phys. Soc. Japan* **13**, 1298–1310.

Ishikawa, Y. and Syono, Y. (1963) Order-disorder transformation and reverse thermoremanence in the $FeTiO_3$–Fe_2O_3 system. *J. Phys. Chem. Solids* **24**, 517–528.

Jacobs, I.S., Beyerlein, R.A., Foner, S. and Remeika, J.P. (1971) Field induced magnetic phase transitions in antiferromagnetic hematite ($\alpha - Fe_2O_3$). *Inter. J. Magnetism* **1**, 193–208.

Lindsley, D.H. (1976) 'Experimental studies of oxide minerals', in *Oxide Minerals* (ed.: D. Rumble III), Mineralogical Society of America, Washington, D.C., Chapter 2.

Merill, R.T. (1968) A possible source for the coercivity of Ilmenite-Hematite minerals. *J. Geomag. Geoelectr.* **20**, 181–185.

Moskowitz, B.M. and Hargraves, R.B. (1982) Magnetic changes accompanying the thermal decomposition of nontronite (in air) and its relevance to Martian mineralogy. *J. Geophys. Res.* **87**, 10115–10128.

Nagata, T. and Akimoto, S. (1956) Magnetic properties of ferromagnetic ilmenites. *Geofis. Pura e Appl.* **34**, 36–50.

Néel, L. and Pauthenet, R. (1952) Etude thermomagnétique d'un monocristal de $Fe_2O_3\alpha$. *C.R. Acad. Sci.(Paris)* **234**, 2172–2174.

Owens, W.H. (1982) A simple model for non-vanishing rotational hysteresis in haematite. *Phys. Earth Planet. Int.* **27**, 106–113.

Schwarz, E.J. (1975) *Magnetic properties of pyrrhotite and their use in applied geology and geophysics*. Geol. Surv. Canada, paper 74–59.

Schwarz, E.J. and Vaughan, D.J. (1972) Magnetic phase relations of pyrrhotite. *J. Geomag. Geoelectr.* **22**, 463–470.

Searle, C.W. and Morrish, A.H. (1966) A three sublattice theory of weakly ferromagnetic $\alpha Me_{\delta}^{4+} Fe_{\delta}^{2+} Fe_{2(1-\delta)}^{3+} O_3$. *J. Appl. Phys.* **37**, 1141–1142.

Smith, R.W. and Fuller, M. (1967) Alpha-hematite: stable remanence and memory. *Science* **156**, 1130 1133.

Soffel, H.C. (1977) Pseudo-single domain effects and the single domain–multidomain transition in natural pyrrhotite deduced from domain structure observations. *J. Geophys.* **42**, 351–359.

Soffel, H.C. (1981) Domain structure of natural fine-grained pyrrhotite in a rock matrix (diabase). *Phys. Earth Planet. Int.* **26**, 98–106.

Stephenson, A. and Collinson, D.W. (1974) Lunar magnetic field palaeo-intensities determined by an anhysteretic remanent magnetization method. *Earth Planet. Sci. Letts.* **23**, 220–228.
Strangway, D.W., Honea, R.M., McMahon, B.E. and Larson, E.E. (1968) The magnetic properties of naturally occurring goethite. *Geophys. J.R. Astr. Soc.* **15**, 345–359.
Syono, Y., Akimoto, S. and Nagata, T. (1962) Remanent magnetization of ferromagnetic single crystal. *J. Geomag. Geoelectr.* **14**, 113–124.
Urquhart, H.M.A. and Goldman, J.E. (1956) Magnetostrictive effects in an antiferromagnetic haematite crystal. *Phys. Rev.* **101**, 1443–1450.
Ward, J.C. (1970) The structure and properties of some iron sulphides. *Rev. Pure & Appl. Chem.* **20**, 175–206.
Westcott–Lewis, M.F. and Parry, L.G. (1971*a*) Magnetism in rhombohedral iron-titanium oxides. *Australian J. Phys.* **24**, 719–734.
Westcott–Lewis, M.F. and Parry, L.G. (1971*b*) Thermoremanence in synthetic rhombohedral iron-titanium oxides. *Australian J. Phys.* **24**, 735–742.
Yamamoto, N. (1968) The shift of the spin flip temperature of α- Fe_2O_3 fine particles. *J. Phys. Soc. Japan* **24**, 23–28.

9 Applications of rock and mineral magnetism

9.1 Introduction

The constitution of rock and mineral magnetism and its 'applications' or 'connections' are discussed in section 1.3. Some applications depend on the physical models of rock and mineral magnetism, and the extent to which rock and mineral magnetism research has provided experimental support for the models. The models essentially concern the acquisition and removal of weak field remanences, and form the physical basis for determinations of the direction and intensity of the ancient geomagnetic field, and the inferences which can be made from such a determination. The models also suggest methods by which the palaeomagnetic signal may be distinguished from 'noise'. The models are applied in geological problems, such as the determination of emplacement temperatures, or in 'magnetic granulometry' with many areas of interest from technology to cosmology.

A second species of application is found in the methods and techniques of rock and mineral magnetism, largely the determination of the composition, concentration and microstructure of the magnetic fraction of a rock or artefact. This provides the chemical basis of palaeomagnetism and palaeointensity studies, and can be used to monitor changes induced in a rock or artefact during laboratory heating. It also finds application as a tool in petrology, in environmental studies, and in the biological/medical area.

The knowledge and experience of rock and mineral magnetism also finds application in a number of other areas which might be grouped together under the heading of 'planetary physics'. They often involve large-scale phenomena, in some cases at remote distances, or the properties of inaccessible materials. Several questions—the properties of the source material contributing to the magnetic field measured at the surface of the Earth or other Earth-type planet, the depth of the Curie point isotherm, the origin of the seismomagnetic and volcanomagnetic effects, the magnetic and magneto-mineralogical effects of meteoritic impact, and the mechanism of magnetization of the crusts of Earth-type bodies in the solar system—may find some answer in the body of rock and mineral magnetism knowledge.

9.2 Physical models—The acquisition and removal of weak field remanences

9.2.1 *Palaeomagnetism*

The processes by which an isotropic magnetic system can preserve a record of the ancient field direction at a particular time, i.e. maintain a non-equilibrium state for the required length of time, are discussed in Chapters 5 and 6. A multicomponent magnetization implies a record of the field at several different times. If the components are carried by mineral fractions with different resistances to thermal and/or AF demagnetization, or if other techniques may be applied (e.g. chemical demagnetization) their resolution becomes a geometrical problem. To assign ages or an order of acquisition to the components cannot be done in isolation, and requires rock magnetic, petrological, geological and other palaeomagnetic information. The papers collected by Soffel and Storetvedt (1978) illustrate the experimental and vectorial procedures adopted in palaeomagnetic studies. Here we consider only the simpler rock magnetic aspects.

As mentioned earlier, each weak field magnetization process has its complementary demagnetization process—the same perturbation being applied in the presence or absence of a field in either case. The ideal procedure for the analysis of the NRM of a rock sample would be to reproduce, in reverse order and in zero field, the sequence of perturbations which occurred between the initial information of the rock and the time at which its NRM was measured. The problem is that the nature, and order of occurrence of the perturbations can only be surmised in many cases, and that the perturbations themselves are generally irreproducible. The conditions under which CRM and DRM were acquired in nature may be complicated or operate over a period which cannot be contracted satisfactorily to the laboratory time scale. A DRM would in any case be the first event in the magnetic history of a sedimentary rock. Depositional demagnetization, the laboratory deposition of crushed, but otherwise unaltered rock in zero field so that it carries zero remanence—a state possibly unachievable by other demagnetization techniques—has applications. A piezoremanence could, in principle, be removed by pressure demagnetization. Chemical demagnetization may be achieved by leaching in acid.

Laboratory analysis of the NRM of rocks almost invariably involves thermal or AF demagnetization regardless of the mechanism by which the NRM may have been acquired. No unequivocal information is therefore obtained about the order in which the revealed components were acquired, unless: (i) all components are carried by a mineral of the same composition in which (ii) the magnetization process is governed by the same intrinsic parameters, and (iii) the remanences have been acquired by the same mechanism. When all three conditions are satisfied, the components would in most (but not all) cases be removed by thermal or AF demagnetization in the reverse order to which they were acquired, and could be meaningfully dubbed 'primary' 'secondary' and so on. A simple case, satisfying all three conditions, consists of a single mineral having a range of grain size but entirely in the monodomain state with barriers to magnetization change resulting from (say) shape anisotropy, and which carries a primary TRM due to the largest grains, a secondary TRM (strictly PTRM) in the middle size range due to reheating ('thermal overprinting') and a VRM in the finest fraction. The relationship between room temperature coercive force and blocking temperature, when the same characteristic time is adopted for both AF and thermal demagnetization, is $H_c(T_0) = H^*(1 - (K_u(T_B)T_0/K_u(T_0)T_B)^{\frac{1}{2}})$, (see, for example, sections 4.4.3 and 6.1.3.) Thus for well-behaved $K_u(T)$, higher blocking temperature corresponds to higher resistance to AF demagnetization.

Numerous cases can be envisaged in which the failure to meet one or more of the conditions results in a differing response to thermal and AF cleaning, or in neither method correctly revealing primary and secondary components. We may take three further simple illustrative cases in each of which one of the above conditions is not met. In the first we might imagine a mixture of (say) magnetite and titanomagnetite, both in the monodomain state with anisotropy due to stress, the magnetite carrying a primary TRM and the titanomagnetite the thermal overprint. Consideration of the relevant properties (section 7.1.3) suggests that AF demagnetization would preferentially remove the primary TRM, whereas thermal demagnetization would preferentially remove the overprint. The second illustration consists of a mineral with a single composition but with differing magnetization process—a fine monodomain fraction with blocking temperature appreciably below the Curie point (carrying the overprint) and a multidomain fraction (carrying the primary component) with large enough barriers to domain wall motion that the unblocking temperature is near the Curie point. At room temperature the coercive force of the monodomain fraction is the higher and the thermal overprint resists AF cleaning. In this, and the two previous cases in which the NRM components are thermally induced (or viscous), the repetition of nature's perturbation during thermal demagnetization successfully reveals the components in the correct sequence. This is not so in our last example, consisting of two generations of the same mineral with the same monodomain magnetization process but formed in different chemical conditions. The first generation carries a primary TRM and the second a CRM, but conditions during grain growth during (say) initial quenching and later low temperature alteration, resulted in the average grain size, and therefore blocking temperature and coercive force, of the second generation mineral being higher than the first. The secondary CRM would then be more resistant to thermal and AF cleaning than the primary TRM. Finally it might be noted that in the event of the first generation fraction also carrying a CRM all three of the above conditions are met, but in this case laboratory thermal or AF demagnetization would not put the components in the correct order of acquisition. The degree of which any individual case satisfies the conditions must be sought from rock magnetic, petrological and geological evidence.

9.2.2 *Palaeointensity studies*

The intensity of a weak field remanence (i.e. in a non-saturating field) is a function of the field, the composition, concentration and microstructure of the magnetic fraction, and the blocking condition. If these latter parameters are constant, or can be accurately reproduced, the intensity is a single valued function of the field. A palaeointensity study consists in experimentally determining this functional dependence by reapplying the natural perturbation and repeating the blocking process in a known field. The NRM intensity then yields the field in which it was acquired.

The difficulties involved are several and various. The only natural perturbation which is likely to allow simple repetition is heating, so that only material carrying a TRM lends itself readily to palaeointensity determination. (That is not to say that other types of remanence have not been considered as potential sources of intensity data). The collected material may have converted part of its original TRM into a VRM, but the contribution of the VRM-carrying fraction can in principle be isolated in both the NRM and laboratory-induced remanences. Any change taking place in composition, concentration and microstructure of the remanence-carrying magnetic mineral fraction since original NRM acquisition, irreversibly distorts or erases the record of the palaeointensity. Laboratory heating itself may also induce physicochemical changes and some monitoring procedure is adopted. It may be possible to determine the intensity from that part of the mineral fraction which remains intact, or from the behaviour before alteration is deemed to have had a significant effect on the remanence-carrying properties of the system.

An anisotropic system does not record a field direction directly (the extreme case is section 5.2.1) and the intensity of acquired remanence will itself be anisotropic. The allowance for this effect in TRM-carrying artefacts is described by Aitken and others (1981). The cooling rate is embodied in the blocking condition—for monodomain grains, the slower the cooling, the lower the effective blocking temperature, and vice versa for multidomain grains. The laboratory cooling rate should therefore match that during NRM acquisition, or (if possible) be allowed for.

Chemical alteration during laboratory heating seems to be a major problem in palaeointensity determinations, even in the (presumably simple) mineralogy of artefacts. One approach has been to substitute the perturbation of elevated temperature by the chemically non-destructive perturbation of alternating field. The NRM lost in a given AF is compared to the

ARM gained in the same AF in the presence of a known direct field. This is analogous to the comparison of thermal demagnetization of NRM to laboratory PTRM acquisition of the heating methods (Fig. 5.4). The ratios of remanences induced in material of a given chemical composition by different perturbations in the presence of the same field are by no means universal, but depend on microstructure and the blocking conditions (sections 6.1.1 and 6.2.1) including the ARM induction method. ARM intensity seems especially dependent on microstructure from the evidence of experimental investigations on magnetite dispersions of varying packing factor (section 7.1.3). The ARM/TRM ratio appropriate to the samples under investigation must somehow be determined in each case. A number of case histories of applications of the ARM method, together with accounts of the methods and problems in palaeointensity determinations, are given in the papers collected by Carmichael (1977).

The models of rock magnetism may have further applications in the quest for paleointensities in providing the basis of other non-heating substitute perturbations. The methods of rock magnetism may continue to play a negative role in providing tests by which particular results may be rejected as unreliable. More positively, the methods may be used in understanding the physicochemical changes which follow from heating as a first step to finding alternative routes to palaeointensity values.

9.2.3 *Geological applications—thermal histories*

As well as recording the direction and intensity of an ambient weak field, the remanence of a system will also record information about the perturbation which led to the remanence being acquired. The perturbation will have interest in geology (sedimentation, diagenetic processes, regional metamorphism, weathering) or in archaeology. A multicomponent remanence records a sequence of perturbations, and laboratory repetition of the blocking conditions allows the perturbations to be studied.

In the case of multicomponent TRM, the thermally overprinted PTRM carries a record of the temperature–time regime experienced during reheating in nature, provided the system has an appropriately wide range of barriers to magnetization change and range of blocking temperatures. Laboratory thermal demagnetization reveals the unblocking temperature, appropriate to the laboratory time scale, i.e. more or less the length of time spent at a particular temperature during the progressive method, of the overprint. If, in nature, the rock had been rapidly heated to an isotherm, held there for the same time as the laboratory thermal demagnetization, and rapidly cooled, the upper limit of the laboratory-determined unblocking temperature spectrum of the reset direction would be the same as that reached during the heating in nature. As this is not likely to be the case, the laboratory unblocking temperature, at which the magnetization reset during a thermal event in nature is lost, must therefore be interpreted in terms of some temperature–time model suggested by geological knowledge. For the simplest model—resetting at an isotherm T_R for time τ_R—the laboratory unblocking temperature T_B appropriate to the characteristic experimental time τ_B is given, by the monodomain model (section 4.4.3) as $T_B = T_R(K_u(T_B) \ln(f_0\tau_R)/K_u(T_R) \ln(f_0\tau_B))$. $K_u(T)$ must therefore be known or assumed to effect the translation from T_B to T_R. Model unblocking curves for magnetite and haematite have been calculated for the isothermal case by Pulliah and others (1975). According to the model calculation, magnetite with shape anisotropy can have (for example) $T_R(10^6\ \mathrm{y}) \sim 200°\mathrm{C}$ and $T_B(\mathrm{ls}) \sim 400°\mathrm{C}$. The next simplest $T(t)$ model corresponds to a constant fall (or rise) in temperature together with a suitable adopted definition of blocking temperature (section 5.1). Dunlop (1981) has considered monodomain magnetite with shape anisotropy, a constant dT/dt, and the 'blocking range' approach to blocking temperature, in which a lower temperature corresponds to a 5% approach to equilibrium and an upper temperature to 95% approach (section 5.1). The result is a set of curves by which the unblocking range under the laboratory cooling rate can be translated to a blocking temperature range

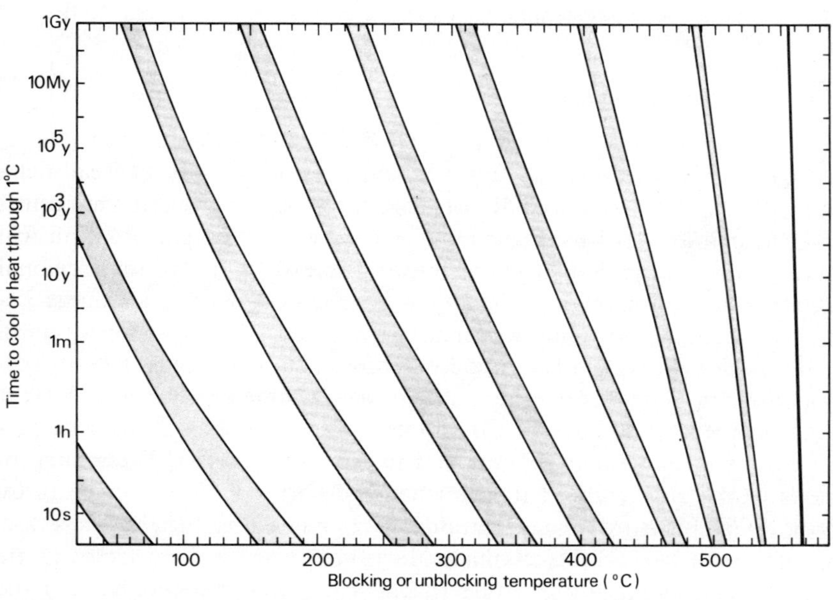

Figure 9.1 The calculated effect of warming or cooling rate on the unblocking or weak-field blocking of monodomain magnetite with shape anisotropy. Each pair of curves and the zone enclosed by them corresponds to particles with the same absolute barrier to magnetization reversal (a constant volume for a given shape) at constant temperature. The region to the left of each zone corresponds to the blocked state. The lower of each pair of curves corresponds to a 5% approach to equilibrium (the demagnetized or remanence-carrying states, depending on the absence or presence of a field) and the upper to 95% approach to equilibrium. (From Dunlop, 1981).

in nature (Fig. 9.1). The relationship between resetting by a thermal pulse and laboratory unblocking temperature has been considered by McLelland-Brown (1981) in a study of contact metamorphism. The emplacement temperature of some South African kimberlites has also been estimated by thermal analysis of multicomponent remanent magnetizations (McFadden, 1977). Studies such as these involve setting up a thermal model for the temperature, as a function of time and position, in the country rock in the neighbourhood of the intrusion. The model depends on the temperature of the intruded rock, the temperature of the country rock before intrusion, the geometry of the intrusion and thermal conductivity. One parameter—the temperature of the intruded material or the pre-intrusion temperature of the country rock—may be left as a variable. The profile of resetting temperature with distance from the intrusion is obtained by translation of the laboratory unblocking temperatures of the reset remanence using the thermal model, The best fit between the model T_{max}—distance profile and resetting temperature profile is found by varying the residual parameter, i.e. the emplacement temperature. The presence of other remanences than TRM, such as CRM in minerals grown during thermal alteration, will of course falsify any conclusion based on thermal analysis and a particular $T(t)$ model, unless such remanence can be identified and isolated.

9.2.4 *Magnetic granulometry*

Intensities of remanence depend on microstructure. In monodomain thermal fluctuation mechanisms the grain volume appears both explicitly and implicitly via the blocking condition. A suitably devised study of the acquisition and removal of weak field remanences by some magnetic system can therefore be interpreted in terms of a particle volume distribution in an assembly of monodomain grains, or a distribution of equivalent effective 'magnetic volumes' in a multidomain assemblage.

The weak field PTRM acquired by a subset of a random array of uniaxial monodomain grains, in a temperature range of some specified width about T_{Bi}, is $\mu_0 F M_s(T_0) M_s(T_{Bi}) w_i v_i H/3kT_{Bi}$ (sections 5.2.1 and 5.2.2). w_i is the fraction of the subset with respect to the whole sample, and the volume range of grains in the subset about v_i has a width depending on the width of the blocking temperature range. If the magnetic material is dispersed in a non-magnetic matrix a further constant fraction F appears as a factor in the PTRMs of all subsets. Grains with $T_B < T_0$ carry no remanence and are thus excluded from an inferred volume distribution. If the composition of the magnetic phase is known, $M_s(T_0)$ and $M_s(T_{BI})$ can be determined and T_{Bi} and H set by the experimenter. A spectrum of $w_i v_i$ can thus be inferred from the histograms of PTRM acquisition or thermal demagnetization of total TRM. The progressive demagnetization of room temperature saturation remanence gives the fraction w_i unblocked in each T_B interval. v_i can then be evaluated.

If the size and temperature dependence of the anisotropy is known (i.e. shape anisotropy), $v_i(T_{Bi})$ may be known directly from the blocking condition. PARMs could also be used, in principle, in magnetic granulometry. High enough AFs would be required to examine the upper end of the monodomain range.

If the grains have the appropriate size energy barriers to magnetization change, the growth or decay of magnetization might be conveniently observed at room temperature on the laboratory time scale. As mentioned in sections 4.4.5 and 6.1.3 the form of the growth and decay curves depends on the relaxation time spectrum. Stephenson (1971*a*) has shown that a model volume distribution of monodomain grains $n(v)dv \propto v^{-2}dv$ results in $\log t$ dependence of VRM acquisition. A volume distribution $n(v)dv \propto v^{-1}dv$ leads to a $\log t$ decay of M_{rs}.

Two examples of magnetically inferred volume distribution of submicroscopic particles in rocks are Stephenson's (1971*b*) investigation of lunar dust and Markert and Steigenberger's (1971) study of a basalt.

Further discussion of the determination of grain sizes of magnetic minerals in rocks from the measurement of magnetization process parameters is given in section 9.3.3.

9.3 Determination of the composition, concentration and microstructure of the magnetic mineral fraction of a rock (or other material)

The concentration of the magnetic minerals in a rock or artefact (or in animal tissue, either in the industrial health or neurological context) is generally small, less than a few percent, and the length scale of the grains can lie below the optical range ($\gtrsim 1\ \mu$m). Sometimes electron optical/analytical methods can be applied successfully in determining the composition, size and shape of the grains (Morgan and Smith, 1981). Separation of the magnetic fraction, using magnetic or other methods, may be successful in producing a sufficiently refined separate that chemical or X-ray analysis can yield useful results. In many cases, however, the magnetic properties of the minerals themselves provide the easiest route to their identification and characterization. This depends on the availability of measured intrinsic properties and magnetization process parameters for well-characterized minerals or synthetic analogues. In some cases the analysis may depend on assuming the applicability of theoretical models, e.g. that for the monodomain magnetization process and the effect of thermal agitations.

The magnetic observables are summarized in Table 9.1. Some depend only on the composition of the contained magnetic mineral fraction, some on composition and concentration, and others on composition, concentration and microstructure. In favourable cases the observables in the last category can be taken together, and the three unknowns determined in a manner analogous to the solution of simultaneous equations. Whereas composition

Table 9.1 The magnetic observables and their dependence on composition, concentration and microstructure (and other extrinsic influences). All depend on composition. Some depend on nothing else (first column). Some have an additional dependence only on microstructure (second column); others only on concentration (third column). Some have additional dependence on both concentration and microstructure (fourth column).

	Intensive, independent of concentration		*Extensive, dependent on concentration, p*	
Dependence on composition ($x, z, y, x_{sulphide}$)	T_c M_F/M_I (and other thermo-magnetic properties)	T_B T_{trans} H_c H_{cr} H_p ratios of extensive properties (M_{rs}/M_s, Q)	M_s χ_{HF}	χ_i W_{rp} M_{rs} (and other remanences)
Dependence on microstructure (*$\bar{L}$, interactions, extrinsic influences*)	—	"	—	"

and concentration can each be specified by a single parameter, microstructure can not. An effective grain size or other more complicated parameter or parameters must be conventionally defined for grains which are not spherical or cubic. Particle-particle interactions and the submicroscopic microstructure and residual stress within the particles, if not taken into account, are also microstructural elements which could falsify an inferred grain size. A single measurement of a microstructure-sensitive property gives an average effective grain size, $\bar{L}$. A series of measurements of some property or properties is needed to determine the grain size distribution $w(L)$ within a sample.

9.3.1 *Composition*

As can be seen from Table 9.1, all measurable parameters depend on composition, although none uniquely define it. This is true even of those properties—T_c, M_F/M_I (see section 7.2.3)—which depend only on composition. A Curie point of 300°C, say, indicates a titanomagnetite with $x \sim 0.4$, or any titanomaghemite composition (x, z) in the quadrilateral (Fig. 2.5) on the 300°C contour, or a haematite-ilmenite solid solution member with $y \sim 0.5$, or monoclinic pyrrhotite ($x_{sulphide} \sim 0.13$). Other thermomagnetic characteristics are semiquantitative indicators of composition. The inversion of a titanomaghemite during a thermomagnetic run (Fig. 7.21) indicates $z > 0$, and the M_F/M_I ratio may prove useful in defining z more precisely than this. More than two composition coordinates may be necessary to usefully specify a titanomagnetite/titanomaghemite (section 2.1.1) and possibly members of the other systems. Other thermomagnetic signatory properties are the γ-transition of low temperature hexagonal pyrrhotites, and magnetic consequences of dehydration of iron oxyhydroxides—the rare non-magnetic lepidocrocite, γ-FeOOH, breaking down to produce strongly magnetic γ-Fe_2O_3.

Apart from the need to conventionally define Curie point temperature when the transition is not sharp, difficulties in observing the Curie point may arise when the concentration of magnetic material is low, and the non-magnetic matrix has an appreciable paramagnetism. In this case the measurement of the temperature dependence of initial susceptibility rather than saturation magnetization may more clearly reveal the Curie point. This will not be the case, however, when the magnetic phase is in the superparamagnetic states as the Curie point is approached (Fig. 4.19). The Curie point temperature of a dilute dispersion of fine grain ferromagnetic particles in a paramagnetic matrix may therefore be difficult to determine.

The anisotropy-related remanence transition temperature, T_{trans}, of magnetite (at the zero point in $K_1 \sim -140°C$) and haematite (at the Morin transition $\sim -10°C$) can be helpful in identifying these minerals. The magnetite transition is lost when other forms of anisotropy dominate (the shape anisotropy of non-spherical monodomain particles), and the haematite transition is grain size dependent.

Values of magnetization process parameters, including the rotational hysteresis parameters, W_{rp} and H_p, can only provide supportive evidence in the identification of magnetic minerals, and then only in the final stages of the iterative process of determination of the composition, concentration and microstructure, when a firm picture of the microstructure may be beginning to emerge.

9.3.2 *Concentration*

The extensive and microstructurally independent properties of Table 9.1 (M_s, χ_{HF}) yield the concentration, once the composition is known. χ_{HF} refers to the high field susceptibility of haematite. When haematite is the only ferromagnetic mineral present and is not superparamagnetic, and other iron-bearing minerals are paramagnets (e.g. clay minerals) a measured χ_{HF} can be combined with chemical analysis for the Fe^{2+} and Fe^{3+} concentrations of the whole rock to give a haematite concentration (Collinson, 1968). In the paramagnetic state the Fe^{2+} (expressed as oxide, FeO) contributes 2.07×10^{-8} m^3/kg to the mass susceptibility, and Fe^{3+}, 2.28×10^{-8} m^3/kg of equivalent non-magnetic oxide, Fe_2O_3. Haematite, αFe_2O_3, contributes 25×10^{-8} m^3/kg after saturation of its ferromagnetism. The chemical analysis provides the FeO and ($Fe_2O_3 + \alpha Fe_2O_3$) concentrations.

The extensive and microstructure-dependent observables provide supportive evidence. If it is believed that the magnetic fraction consists of multidomain spheres of some strongly magnetic, soft material (e.g. magnetite, titanomagnetite, iron) with volume susceptibility 1/3, the concentration may be obtained without direct knowledge of the composition. If the magnetic phase is in the form of fine randomly orientated needle-like grains, and believed to be in the monodomain state with shape anisotropy, M_{rs} measurement may be substituted for M_s ($M_{rs}/M_s = 0.5$).

9.3.3 *Microstructure*

The intensive magnetization process parameters and transition temperatures depend only on microstructure once composition has been determined, as do ratios of extensive properties (Table 9.1) such as M_{rs}/M_s and the Königsberger ratio, Q (= weak-field remanence per unit field divided by susceptibility). Magnetic measurements provide many clues to the microstructure. There are several elements of microstructure, however—the size and shape of grains, the geometrical relationship between grains, the orientation of grains, submicroscopic microstructure (voids, inclusions, the local state of strain)—and some elements may need to be described by several conventionally defined parameters. The difficulty in specifying the size and shape of irregular particles has already been mentioned. Of the three factors which are to be determined in reaching a description of the magnetic mineral content of a rock or artefact, the microstructure will probably be the least well defined in most cases.

9.3.3.1 *The grain size.* Here we are concerned with the determination of grain size and not necessarily the domain state of the minerals. The nature of the domain state is the goal of rock magnetism studies in which the physical basis of the observed magnetization process and properties of weak field remanences is being sought. In palaeomagnetism, what is important is the empirically determined stabilities of remanences rather than the physical origins of stability.

A first estimate of average or characteristic sizes of mineral grains in a rock or other sample can be obtained, once composition is known, by comparing measured values of the intensive microstructure-dependent observables of the right-hand side of Table 9.1 with data like that shown in Figs. 7.9, 7.10, 7.11, 8.1, 8.5 and 8.6. Concentration may also be known, in which case the extensive properties can also be exploited by comparison with Figs. 7.8 and possibly 7.16 and 7.17. In a suite of samples containing a single mineral, trends in intensive properties or ratios indicate microstructural trends, presumably grain size trends in by far the majority of cases. The success of such determinations of average grain size depends on how well the factors governing the magnetization process in the synthetic analogues or minerals, providing the reference values, coincide with those of the specimen under investigation. In some cases, e.g. the pyrrhotites, reference data are unavailable.

If the mechanism controlling the magnetization process is assumed, a grain size may be inferred using the relevant model. For aligned monodomain grains with uniaxial anisotropy, the temperature dependence of coercive force is $H_c(T) = H^*(T)(1 - (\ln(f_0\tau)kT/K_u(T)v)^{\frac{1}{2}})$ where $H^*(T) = 2K_u(T)/\mu_0 M_s(T)$. When anisotropy is due to shape, $K_u(T) \equiv (1/2)\mu_0(N_b - N_a)M_s^2(T)$ for ellipsoids of revolution. Substitution and rearrangement yields

$$\frac{H_c(T)}{M_s(T)} = (N_b - N_a) - \left(\frac{2\ln(f_0\tau)(N_b - N_a)}{\mu_0 v}\right)^{\frac{1}{2}} \cdot \frac{T^{\frac{1}{2}}}{M_s(T)}.$$

This expression is the basis of Dunlop's (1976) 'thermal fluctuation analysis'. The determination of $M_s(T)$ and $H_c(T)$, and a straight line fit to the plot of $H_c(T)/M_s(T)$ against $T^{\frac{1}{2}}/M_s(T)$, gives $(N_b - N_a)$ from the intercept and v from the slope. Where a range of grain sizes is present, the analysis yields a mean $(N_b - N_a)$ and mean $v^{\frac{1}{2}}$. The temperature range for H_c measurement must not extend beyond the blocking temperature of the finest material, the presence of superparamagnetic grains invalidating the above expression. The measurement of $H_{cr}(T)$ avoids this difficulty.

Methods of 'magnetic granulometry', in which the particle size distribution is determined, are based on models for the thermal activation of the moments of monodomain grains (as in the weak-field remanence model application of section 9.2.4) or, empirically, after experimentally resolving the contribution of different grain size fractions to the magnetization process parameters (principally determination of the coercivity spectrum). In the former category are the granulometry studies of Weil, e.g. Weil (1949), in which the temperature dependence of H_c and M_{rs}/M_s is determined for fine particles over the blocking range. At a temperature above the highest blocking temperature, at which the assemblage is entirely in the superparamagnetic state, the magnetization curve for isotropic grains is $M = \mathscr{L}(\alpha)$, the Langevin function with $\alpha \equiv \mu_0 M_s Hv/kT$ (section 4.4.4). If M_s is independent of v (not necessarily so in ultrafine particles) and known, a Langevin function can be fitted to the magnetization curve with v as the variable parameter. When a distribution of grain sizes occurs v is replaced by $\sum w_i v_i$ and a unique solution to the problem cannot be obtained. An indication of the range of particle sizes can be got by comparing the low field part of the curve (governed by the larger particles) and the approach to saturation (dominated by the smaller particles) (see Jacobs and Bean, 1963). The variation of initial susceptibility with temperature and frequency for a mixture of monodomain and superparamagnetic grains, in which the monodomain susceptibility is independent of temperature, and in which the grain size distribution is given by an analytic function of the type $n(v) \propto v^p$, has been derived by Stephenson (1971*a*).

The coercivity spectrum is also a spectrum of the microstructural influence on the magnetization process, when a single mineral is present. The coercivity spectrum may be produced by the progressive direct field demagnetization of M_{rs}, and presented as a histogram of incremental loss of remanence in each field interval. A grain size interval is assigned to each field interval and the quantity of material in each interval inferred from the incremental remanence loss. This, of course, depends on the availability of data for the size dependence of H_{cr} and M_{rs}. The coercivity spectrum can also be obtained from AF demagnetization of remanence and progressive acquisition of isothermal remanence. The rotational hysteresis (W_R–H) curve (section 4.3.5) is also a type of coercivity spectrum.

9.3.3.2 *Fabric.* The magnetization process parameters of a single grain or an aligned assemblage of identical grains will be anisotropic except in special cases.

Taking first the monodomain model of section 4.3.1, the susceptibility, per unit volume of magnetic material, in an aligned assemblage of grains is $\mu_0 M_s^2 \sin^2 \phi / 2K_u$ where ϕ is the angle between the field and the unique axis of the system. $M_{rs}(\phi) = M_s \cos \phi$, and the coercive force is a function of ϕ and K_u (Fig. 4.13) as is the constant in the law of approach to saturation (section 4.3.1). For monodomain ellipsoids of revolution with dominant shape anisotropy $K_u \equiv (1/2)\mu_0(N_b - N_a)M_s^2(T)$.

The model multidomain grain, with low resistance to domain wall motion, has susceptibility $1/N$ parallel to the domain magnetization and a monodomain-like perpendicular susceptibility. For the complicated case of many non-linear walls, the susceptibility is commonly taken to be simply $1/N$. An ellipsoidal grain or an aligned assembly of grains has principal susceptibilities, per unit volume of magnetic material, $\chi_a \simeq 1/N_a$ and $\chi_b \simeq 1/N_b$. The model coercive force is independent of direction (section 4.3.3.2) but $M_{rs} \simeq H_c/N$ and H_{cr} may also be direction-dependent. We might note that the assemblage of aligned elongated grains has maximum susceptibility along the stretched axis when the grains are multidomain but a minimum when monodomain.

Model assemblages intermediate between the aligned and randomly dispersed cases could be posited and the corresponding anisotropy in the magnetization process parameters calculated. Real assemblages have apparently reasonably simple and symmetrical distributions of grain preferred axes, as measurements of the anisotropy of susceptibility can be satisfactorily reduced to three principal susceptibilities which are, in effect, statistics of the distribution. Further parameters derived from the three principal susceptibilities characterize the distribution in terms of 'lineation', 'foliation' and 'eccentricity'. Collinson (1983) has reviewed and compared methods for measuring the anisotropy of susceptibility, and an example of the application of the technique to the study of strain in deformed sediments is the work of Kligfield and others (1981).

9.3.3.3 *Interactions.* The overall concentration can be determined using the extensive microstructure-independent properties of Table 9.1. Local concentrations will be lower and higher than the overall value. If the magnetic particles are mainly localized in clusters, for whatever reason, the effective concentration—that important in the microstructural context—is higher than the overall concentration. This may be detectable when the close proximity of magnetic particles leads to magnetostatic interactions.

Magnetostatic interactions between particles have been widely invoked to account for the differences between experimental observation and monodomain model values of the magnetization process parameters, and the model characteristics of weak field remanence acquisition. An example of this is the 'unshearing' of observed hysteresis loops of material believed to to be in the monodomain state with uniaxial anisotropy, so that M_{rs}/M_s translates from the measured value to the model value of 0.5 (Davis and Evans, 1976). The unshearing operation yields a value of pN where p is the local packing factor and N the effective demagnetizing factor of the cluster (estimated from microscopic examination or simply taken as 1/3).

Another example involves the complementarity of the magnetization and demagnetization processes. A system magnetized to saturation by application of a field $+H_s$, and the field removed, has remanence vectors in one hemisphere if monodomain, or an equivalent distribution of domain directions if multidomain. The application and removal of a reverse field, $-H$, leaves grains with switching field $<|H|$ in the negative hemisphere with the corresponding loss of vectors in the positive hemisphere (like Fig. 4.13). The negative hemisphere has a remanent moment of $M_r(H)$ and the positive hemisphere has a moment $M_{rs} - M_r(H)$. The moment of the (partially) demagnetized assemblage, $M_D(H)$, is then $M_{rs} - 2M_r(H)$. If we start with a randomly magnetized assemblage, i.e. after AF demagnetization, and apply and remove a field of $\pm H$, the resultant moment is the remanence of those grains with switching field $<|H|$, i.e. $M_r(H)$. In two separate experiments $M_r(H)$ and $M_D(H)$ can be determined and the applicability of the relationship $M_D(H) = M_{rs} - 2M_r(H)$ verified. Bate (1962) ascribed the failure of the relationship in experiments on γ-Fe_2O_3 and cobalt-substituted γ-Fe_2O_3 to interactions between the particles. The discrepancy was smaller in the cobalt-substituted material in which anisotropy was not believed to be dominated by shape. Similar experiments (but using AF demagnetization) on a plagioclase with dispersed magnetite inclusions, a diabase (a titanomagnetite-containing intrusive), 'biomagnetic' magnetite, and dispersed synthetic magnetite are reported by Cisowski (1981). The results seem to correlate well with the expectation of interactions, the dispersed synthetic (with overall concentration only $\sim 5\%$, but presumably high local concentrations) being the most affected.

In more complicated magnetization/demagnetization experiments, in which sequences $(+H_s, -H_1, +H_1)$, $(+H_s, -H_2, +H_1, +H_2)$, $(+H_s, -H_3, +H_1, +H_2, +H_3)$ etc. are applied, a two-dimensional coercivity spectrum may be derived, where the two dimensions are the coercive forces in the + and − directions (a Preisach diagram). The soundness of the theoretical basis for the Preisach diagram may be debatable. The diagram is nevertheless, an interesting and possibly useful tool. The determination and display of the diagram presents an ideal and simple problem of the kind which lends itself readily to solution using a microprocessor-controlled instrument and VDU screen display. For details of how to construct a Preisach diagram, and for photographs of models of the resultant three-dimensional histograms, the reader is referred to Bate (1962).

Dunlop (1969) determined the Preisach diagram for, among other material, a basalt sample with deuterically oxidized titanomagnetite—the magnetic constituent of the resultant intergrowths being expected to experience interaction fields. The off-diagonal parts of the histogram, for which grains have different $+H_c$ and $-H_c$ coercive forces because of local interaction fields, were used to modify the Néel monodomain TRM model for uniaxial aligned grains (section 5.2.1), so that experimentally observed TRM acquisition curves and model were in agreement. The presence of interactions need not be invoked, however, if the experimental observations are compared to the TRM model with randomly dispersed grain axes (section 5.2.2) (Dunlop, 1981). The Preisach diagram may be helpful in elucidating the possibly more important role of interactions in ARM acquisition.

9.4 Connections in planetary physics—the magnetization of planetary crusts

We are not concerned here with studies of the remanent magnetization of meteorites or lunar samples. The rock magnetic or palaeomagnetic aspects of such studies are probably no different from equivalent studies on terrestrial rocks. The magnetic minerals are exotic but simple by comparison with terrestrial magnetic mineralogy. The remanence acquisition mechanisms—TRM, CRM and PRM—of terrestial rocks may also apply, although remanence induced by shock, SRM, has been postulated in this context. The source of the ancient inducing field is another matter and the importance of lunar samples and meteorites in extending the horizons of the planetary scientist, not only in space but also in time—extending to the origin of the solar system itself—cannot be underestimated. Here we are concerned with speculation on or model building of large-scale phenomena, or the properties of inaccessible material.

In palaeomagnetic studies the field due to small samples of the Earth's crust, collected from near the surface, are measured using a suitable magnetometer. Large-scale magnetic surveys, that use a magnetometer carried by a satellite, aircraft, surface vehicle or submersible, measure the field due to large samples of the Earth's crust, the dimensions of the sample depending on the altitude of the magnetometer. The main field due to the Earth's core has first to be defined and subtracted from the measured field to leave the 'anomaly' due to crustal magnetization. While it is possible, at least in principle, to analyse the source of the magnetism of the small sample, in the remanence measuring magnetometer, in terms of the physics and chemistry of the constituent minerals, it may not be possible to explain the field due to large samples of the crust because their scale prohibits representative sampling and, in any event, the material contributing to the anomaly may lie as deep as several km. The problem is the same with the large crustal or small palaeomagnetic sample—the composition, concentration, microstructure and probable remanence-inducing mechanism is required in either case. The small sample is measured in zero field in the remanence measuring magnetometer. This is not so for the large sample measured by the surveying magnetometer, and an induced magnetization due to the ambient field and the susceptibility of the sample is also present. One question is the size of the crustal sample. The lateral dimensions are related to the extent or wavelength of the anomaly; the third dimension, the depth, is unknown. The layering of the crust means that several superposed samples are contributing to the anomaly. Each mineral has a Curie point isotherm—the depth at which the geothermal gradient removes the spontaneous magnetization, and any significant contribution to the overlying anomaly, of that species. The depth and hence volume of the crustal sample therefore depends on both mineral species and the geothermal gradient. Variations in the sample volume add an extra unknown not present in the problem of explaining the magnetism of a palaeomagnetic sample.

To the questions to be asked in a rock magnetic study of palaeomagnetic material—the magnetic mineralogy, the mechanism, the effect of alteration (serpentinization, maghemitization) on remanence, thermal or viscous overprinting—are added further complications of mechanical disturbance in the sample, the fact that the sample has a non-uniform temperature, the possibility of exotic metallic minerals with high Curie point (Haggerty, 1978) and an induced magnetic moment. To make the problem more interesting, the scientist must operate with the handicap of being able to perform no experiments on the sample. Fortunately scientists are not easily discouraged. Harrison (1983) reviews some of the recent literature.

Magnetic surveys have also been carried out over part of the lunar surface and some information about the magnetic fields in the vicinity of other Earth-like planets is also available. In the case of the Moon the whole of the observed field is the anomaly, as there is believed to be negligible field of internal origin. This is not necessarily so for the planet Mercury. The thermal and chemical histories of other planets are less well established so that the applicability of the TRM and CRM mechanisms, which are proba
so important in the Earth's crust, is uncertain. In one lunar sc
accreting core acquires a DRM, IRM or VRM in a prim
field. A hot outer shell may acquire TRM while cooling i
this magnetized core. The magnetized core later loses its r

temperature rises. Magnetostatic theorems for magnetized shells and magnetostatic interactions, developed in the mineral magnetism context to describe inclusions of one magnetic phase within another, but now applied to planetary size bodies, may be important. Meteoritic impact may play an important role in modifying the weak field external to the magnetized shell through excavation of the shell, by imparting a shock remanent magnetization (SRM) locally and to ejected material, by producing a transitory locally enhanced field, and possibly by physico-chemical processes occurring at the antipodes to the impact site. Reviews of the possible magnetization mechanisms and origins of the magnetizing field are to be found in Daily and Dyal (1979) and Hood and Cisowski (1983).

Stress modifies the anisotropy via magnetostriction (section 3.5). The positions of the preferred axes of monodomain grains and directions of domain magnetization may be significantly modified by uniaxial stress. The properties of some materials (e.g. fine grain ball-milled titanomagnetites) can be dominated by the residual stress in the grains. The magnetization process parameters (and the superparamagnetic/stable monodomain boundary) and directions and intensities of remanences may depend on the stress, given a suitable combination of magnetostriction constants and large enough stress, so that contributions from shape or crystal anisotropy are not dominant. Theoretical models for the dependence of magnetic properties on stress are developed by Nagata (1970) and Stacey and Johnston (1972). The remanent and induced magnetization of crustal rocks is therefore, in principle at least, sensitive to the changes in stress experienced during tectonic processes. The time dependence of the magnetic field due to crustal magnetization contains some contribution from the build-up and relaxation of stress in the crust (tectonomagnetism), and the search for the seismomagnetic effect, with its prospects for earthquake prediction, continues. The existence of the volcanomagnetic effect seems better established (Davis, 1983).

9.5 Summary

Models for the acquisition of weak field remanence describe how the direction of the field is recorded by a magnetic system at the time at which a remanence-inducing perturbation is applied. The remanence also carries information about the intensity of the field, the nature of the perturbation and the microstructure of the magnetic grains. The models suggest ways by which the information may be retrieved—the essential operation being the laboratory repetition of the perturbation itself. The perturbation most likely to lend itself to successful repetition in the laboratory is the thermal agitation of TRM and VRM, and laboratory heating can lead to analysis of multicomponent NRM, determination of palaeointensities, information about thermal history and magnetic granulometry. Reliable results require that no chemical change takes place in the laboratory and that the NRMs are TRMs or VRMs.

The magnetic observable—magnetization process parameters, remanences, Curie point temperatures, thermomagnetic characteristics, transition temperatures (table 9.1)—depend on the composition, concentration and microstructure of the magnetic fraction of the sample under investigation. The composition, concentration and microstructure can therefore be determined by measurement of the observables, together with reference data for the appropriate magnetic materials and/or the application of suitable models for the magnetization process.

The constitution of the magnetization of the large parts of a planetary crust, which give rise to the anomalies in the magnetic field measured near the surface of the planet, is difficult to determine, but the problems are essentially the same as those faced in a 'rock magnetic' investigation of a 'palaeomagnetic' sample. Because the problem concerns at present unknowable data, the answers based on rock and mineral magnetism knowledge are not definitive and often only speculative. No doubt advancing technology and the intensive interest in magnetic anomalies arising from their importance in understanding the Earth's crust will significantly reduce the present uncertainties.

Bibliography

Aitken, M.J., Alcock, P.A., Bussell, G.D. and Shaw, C.J. (1981) Archaeomagnetic determination of the past geomagnetic intensity using ancient ceramics: allowance for anisotropy. *Archaeometry* **23**, 53–64.

Bate, G. (1961) Statistical stability of the Preisach diagram for particles of γFe_2O_3. *J. Appl. Phys.* **33**, 2263–2269.

Carmichael, C.M. (ed.) (1977) Palaeomagnetic field intensity, its measurement in theory and practice. *Phys. Earth Planet. Int.* **13**, 241–390.

Cisowski, S. (1981) Interacting vs. non-interacting single domain behaviour in natural and synthetic samples. *Phys. Earth Planet. Int.* **26**, 56–62.

Collinson, D.W. (1968) An estimate of the haematite content of sediments by magnetic analysis. *Earth Planet. Sci. Letts.* **4**, 417–420.

Collinson, D.W. (1983) *Methods in Rock Magnetism and Palaeomagnetism: Techniques and Instrumentation.* Chapman & Hall, London, New York.

Daily, W.D. and Dyal, P. (1979) Theories for the origin of lunar magnetism. *Phys. Earth Planet. Int.* **20**, 255–270.

Dunlop, D.J. (1969) Preisach diagrams and remanent properties of interacting monodomain grains. *Phil. Mag.* **19**, 369–378.

Dunlop, D.J. (1976) Thermal fluctuation analysis: a new technique in rock magnetism. *J. Geophys. Res.* **81**, 3511–3517.

Dunlop, D.J. (1981) The rock magnetism of fine particles. *Phys. Earth Planet. Int.* **26**, 1–26.

Davis, P.M. (1983) Tectonomagnetism. *Rev. Geophys. Space Phys.* **21**, 685–693.

Davis, P.M. and Evans, M.E. (1976) Interacting single-domain properties of magnetite intergrowths. *J. Geophys. Res.* **81**, 989–994.

Haggerty, S.E. (1978) Mineralogical constrains on Curie isotherms in deep crustal magnetic anomalies. *Geophys. Res. Letts.* **5**, 105–108.

Harrison, C.G.A. (1983) Magnetic anomalies. *Rev. Geophys. Space Phys.* **21**, 634–643.

Hood, L.L. and Cisowski, S.M. (1983) Paleomagnetism of the Moon and meteorites. *Rev. Geophys. Space Phys.* **21**, 676–684.

Jacobs, I.S. and Bean, C.P. (1963) 'Fine particles, thin films and exchange anisotropy (effects of finite dimensions and interfaces on the basic properties of ferromagnets)', in *Magnetism*, (ed. G.T. Rado and H. Suhl) Academic Press, New York, pp. 271–350.

Kligfield, R., Owens, W.H. and Lowrie, W. (1981) Magnetic susceptibility anisotropy, strain and progressive deformation in Permian sediments from the Maritime Alps (France). *Earth Planet. Sci. Letts.* **55**, 181–189.

Markert, H. and Steigenberger, N. (1971) On the size distribution of submicroscopic magnetite and titanomagnetite fine particles in basalt. *J. Geophys.* **37**, 499–518.

McLelland-Brown, E. (1981) Paleomagnetic estimates of temperatures reached in contact metamorphism. *Geology* **9**, 112–116.

McFadden, P.L. (1977) A palaeomagnetic determination of the emplacement temperature of some South African kimberlites. *Geophys. J.R. astr. Soc.* **50**, 587–604.

Morgan, G.E., and Smith, P.P.K. (1981) Transmission electron microscope and rock magnetic investigations of remanence carriers in a precambrian dolerite. *Earth Planet. Sci. Lett.* **53**, 226–240.

Nagata, T. (1970) Basic magnetic properties of rocks under the effect of mechanical stresses. *Tectonophysics* **9**, 167–195.

Pulliah, G., Irving, E., Buchan, K.L. and Dunlop, D.J. (1975) Magnetization changes caused by burial and uplift. *Earth Planet. Sci. Letts.* **28**, 133–143.

Soffel, H. and Storetvedt, K.M. (eds.) (1978) Reliability of palaeomagnetic results: criteria, methods and error estimation. *Phys. Earth Planet. Int.* **16**, 1–83.

Stephenson, A. (1971*a*) Single domain grain distributions. I. A method for the determination of single domain grain distributions. *Phys. Earth Planet. Int.* **4**, 353–360.

Stephenson, A. (1971*b*) Single domain grain distributions. II. The distribution of single domain iron grains in Apollo 11 lunar dust. *Phys. Earth Planet. Int.* **4**, 361–369.

Weil, L. (1949) Champ coercitif et granulometrie du Nickel Raney, *C.R. Acad. Sci.* (*Paris*) **229**, 584–585.

Index